Matrix Methods in Stability Theory

Applications of Mathematics Series

Editor: Alan Jeffrey
Professor of Engineering Mathematics
University of Newcastle upon Tyne

WILLIAM F. AMES
Numerical Methods for Partial Differential Equations

S. BARNETT and C. STOREY
Matrix Methods in Stability Theory

T. J. M. BOYD and J. J. SANDERSON
Plasma Dynamics

C. D. GREEN
Integral Equation Methods

I. H. HALL
Deformation of Solids

JEREMY HIRSCHHORN
Dynamics of Machinery

ALAN JEFFREY
Mathematics for Engineers and Scientists

BRIAN PORTER
Synthesis of Dynamical Systems

Matrix Methods in Stability Theory

S. BARNETT
University of Bradford

C. STOREY
University of Technology, Loughborough

Barnes & Noble, Inc.
New York
Publishers • Booksellers • Since 1873

First published in Great Britain 1970 by
THOMAS NELSON AND SONS LTD

First published in the United States of America 1970 by
BARNES & NOBLE, INC.

Printed in Great Britain

Contents

Preface

In the past two decades an increasing amount of attention has been paid to theoretical problems arising from the control of engineering and other systems. This has involved the development of a number of mathematical techniques, many of which break away from the 'classical' aspects of applied mathematics. The aim of this book is to introduce young graduate or undergraduate mathematicians and mathematically inclined engineers and scientists to certain aspects of matrix algebra which arise through the use of Liapunov's stability theory of ordinary differential equations. We also hope that it will be useful to those already working in control theory. The key to the development of the book is the matrix equation which arises when a quadratic form is used as a Liapunov function for a system of linear equations. Since the solution of this equation resolves the stability problem for the system it is obviously connected in a fundamental way with the characteristic roots of the system matrix. It is this connection which allows results concerning the nature of the characteristic roots to be obtained in novel, powerful, and, in our opinion, interesting ways with remarkable ease. Furthermore it is possible to deduce general results concerning the effect of arbitrary perturbations on the stability and optimality of both linear and non-linear systems.

The book is not meant to be a text on matrix algebra nor a descriptive account of Liapunov theory and so far as these subjects are concerned we have simply stated the results we need as theorems without proof and referred the reader to the many excellent existing texts for further details.

We have, however, included a chapter on special forms of matrix and the connection between them since these are particularly important for the rest of the book. In particular the companion matrix, the Schwarz matrix, the Jacobi matrix and the Jordan and Frobenius forms are dealt with. The remainder of the book deals with the solution of the Liapunov matrix equation and its use in the areas mentioned above. This work has been the subject of our own researches and is dealt with in some detail although we have tried as much as possible to avoid tedious routine algebraic manipulation. We have covered the relevant research literature without attempting to be exhaustive and only made brief reference to the work of others where it has directly impinged on our own.

The most important prerequisite for the book is a good working knowledge of matrix algebra and we have also assumed familiarity with general basic analytical techniques. The book has been consciously kept as elementary as possible in order to emphasize the power and simplicity of the techniques.

Finally, the book can be used in two possible ways. The first is to give some interesting non-classical applications of matrix theory and to show connections between a number of seemingly dissimilar parts of mathematics at the undergraduate level. The second is to indicate a number of areas in which both applied mathematicians and control theorists could begin research. In addition we hope that the concise statements of theorems and definitions will prove useful for purposes of reference.

Chapters 1, 2, 7, 8, and 9 were written by the first-named author and Chapters 3, 4, 5, and 6 by the second.

We would like to thank Mr. G. T. Joyce for reading and commenting on the manuscript and Mrs June Russell for her skilful typing.

S.B.
C.S.

1

Matrix algebra : introduction

1–1 Matrices and vectors

A *matrix* A of *order* $m \times n$ can be regarded as an array of mn numbers, or *elements*, in m rows and n columns. If the element in the ith row and jth column is denoted by a_{ij} then it is customary to write $A = (a_{ij})$. The elements generally belong to the field of complex numbers **C**, although it is usually the case in problems arising from dynamical systems that they are drawn from the field of real numbers **R**. A matrix with elements in a field **F** is said to be *over* **F**. When $m = 1$ the matrix reduces to a *row* n-*vector* and when $n = 1$ to a *column* m-*vector*; the elements of a vector are usually called *components*. In the case when $m = n$, A is called a *square* matrix of order n. A special form of square matrix is when $a_{ij} = 0$, $i \neq j$; such a matrix is called *diagonal* and written $A = \text{diag}\,(a_{11}, a_{22}, \ldots, a_{nn})$. In particular the nth-order *unit matrix* $I_n = (\delta_{ij})$ is defined by $I_n = \text{diag}\,(1, 1, \ldots, 1)$. For any square matrix the elements $a_{11}, a_{22}, \ldots, a_{nn}$ form the *principal* (or *leading*) diagonal of A.

An *upper* (*lower*) *triangular* matrix $A = (a_{ij})$ is a square matrix all of whose elements below (above) the principal diagonal are equal to zero. That is, $a_{ij} = 0$, $i > j$ $(i < j)$.

A *zero* (*or null*) *matrix* is a matrix having every element zero.

1–2 Basic operations

Two matrices $A = (a_{ij})$ and $B = (b_{ij})$ are said to be *equal* if and only if they are of the same order and $a_{ij} = b_{ij}$ for each pair of suffixes i and j. The *sum* of A and B is defined as the matrix whose (i,j) element is $a_{ij} + b_{ij}$.

The product of A and a scalar k (drawn from the same field as the elements of A) is defined simply as the matrix whose (i,j) element is ka_{ij}. The product $C = AB$ of two matrices $A(m \times n)$ and $B(l \times p)$ is defined only if $n = l$ and is then the $m \times p$ matrix (c_{ij}) where

$$c_{ij} = \sum_{k=1}^{n} a_{ik}b_{kj},\ i = 1, \ldots, m;\ j = 1, \ldots, p,$$

A and B then being said to be *conformable for multiplication*. B is said to be *pre-multiplied* by A, and A *post-multiplied* by B. Clearly, from the definitions, addition and multiplication of matrices are associative:

$$(A + B) + C = A + (B + C);\ A(BC) = (AB)C$$

and distributive:

$$A(B + C) = AB + AC.$$

Addition is commutative:

$$A + B = B + A,$$

but multiplication generally is not:

$$AB \neq BA.$$

For example, if $a = (a_i)$ is a row n-vector and $b = (b_i)$ a column n-vector then $ab = \sum_{i=1}^{n} a_i b_i$, the *scalar product* of the two vectors, but ba is an $n \times n$ matrix. If AB does equal BA then the matrices A and B are said to *commute*.

The rth *power* of a square matrix A is defined by $A^r = A \,.\, A^{r-1}$, $r = 2, 3, \ldots$. Clearly all powers of A commute with each other.

1–3 Transpose

$A = (a_{ij})$ is an $m \times n$ matrix then the *transpose* of A, written A' (or A^T), is the $n \times m$ matrix whose (i, j) element is a_{ji}. When the elements of A belong to **C** the *conjugate transpose of A* is defined by $A^* = (\bar{a}_{ji})$. Thus if $\bar{A} = (\bar{a}_{ij})$ is the *conjugate* of A then $A^* = (\bar{A})'$. For real matrices (i.e., over **R**) A' and A^* are the same.

Theorem 1–3–1

The transpose of a product of matrices is equal to the product of the transposed matrices but with the order of the factors reversed; that is, $(AB)' = B'A'$. Similarly $(AB)^* = B^*A^*$.

Also it follows very easily that $(A + B)^* = A^* + B^*$; $(kA)^* = \bar{k}A^*$; and $(A^*)^* = A$.

1–4 Some special matrices

Some matrices which are of great importance are now defined.

A matrix A is said to be *normal* if

$$A^*A = AA^*.$$

Examples of normal matrices are diagonal matrices; *hermitian* matrices, defined by

$$A^* = A;$$

skew-hermitian matrices, defined by

$$A^* = -A;$$

and *unitary* matrices, defined by

$$UU^* = U^*U = I.$$

If the matrices have real elements then hermitian, skew-hermitian, and unitary matrices are called *symmetric*, *skew-symmetric*, and *orthogonal* respectively.

In particular the elements on the principal diagonal of a hermitian matrix are all real since $\bar{a}_{ii} = a_{ii}$, of a skew-hermitian matrix are all pure imaginary since $\bar{a}_{ii} = -a_{ii}$, and of a skew-symmetric matrix are all zero.

Exercise 1–4–1
Show that the product of two unitary matrices is again unitary.

Exercise 1–4–2
Show that if $AA^* = 0$ then $A = 0$. (A is an $m \times n$ complex matrix.)

1–5 Partitioning

Let A be an $m \times n$ matrix and $k \leqslant m, l \leqslant n$. Then any k rows and l columns of A determine a $k \times l$ *submatrix* of A. It is convenient to write

$$A(i_1, i_2, \ldots, i_k \mid j_1, j_2, \ldots, j_l)$$

for the submatrix formed by selecting rows $i_1, \ldots, i_k$ and columns $j_1, \ldots, j_l$ of A.

Partition A into submatrices α_{ij} of order $m_i \times n_j$ as follows:

$$A = \begin{array}{c} \begin{array}{cccc} n_1 & n_2 & \cdots & n_s \end{array} \\ \left[\begin{array}{c|c|c|c} \alpha_{11} & \alpha_{12} & \cdots & \alpha_{1s} \\ \hline \alpha_{21} & \alpha_{22} & \cdots & \alpha_{2s} \\ \hline \cdot & \cdot & \cdots & \cdot \\ \cdot & \cdot & \cdots & \cdot \\ \cdot & \cdot & \cdots & \cdot \\ \hline \alpha_{r1} & \alpha_{r2} & \cdots & \alpha_{rs} \end{array}\right] \begin{array}{c} m_1 \\ m_2 \\ \cdot \\ \cdot \\ \cdot \\ m_r \end{array} \end{array}$$

where

$$m_1 + m_2 + \cdots + m_r = m, \qquad n_1 + n_2 + \cdots + n_s = n.$$

Similarly, partition the $n \times p$ matrix B into submatrices β_{ij} of order $n_i \times p_j$ where $p_1 + p_2 + \cdots + p_t = p$.
Then

Theorem 1–5–1
$AB = C$ where C is partitioned into submatrices γ_{ij} of order $m_i \times p_j$

and
$$\gamma_{ij} = \sum_{k=1}^{s} \alpha_{ik}\beta_{kj}.$$

That is, provided A and B are partitioned conformably into submatrices their product AB can be evaluated by regarding the submatrices exactly as though they were scalar elements and applying the usual rule for multiplication.

As an example of partitioning, the *direct sum* of the square matrices A, B of orders n, p respectively is defined by

$$A \oplus B = \begin{array}{c} \begin{array}{cc} n & p \end{array} \\ \left[\begin{array}{c|c} A & 0_{np} \\ \hline 0_{pn} & B \end{array}\right] \end{array} \begin{array}{c} n \\ p \end{array}$$

where 0 denotes zero matrices of appropriate order. Clearly the operation $\oplus$ is associative but not commutative. Generalizing, $\sum_{i=1}^{k} \oplus A_i$ denotes the sum $A_1 \oplus A_2 \oplus + \cdots \oplus A_k$ and is often called a *quasi-diagonal* matrix, and written diag $(A_1, A_2, \ldots, A_k)$. It follows at once that a special case of Theorem 1–5–1 is

$$(\sum_{i=1}^{k} \oplus A_i)(\sum_{i=1}^{k} \oplus B_i) = \sum_{i=1}^{k} \oplus A_i B_i,$$

where A_i and B_i are of the same order n_i.

1–6 Kronecker product

Another useful way of combining matrices is by forming their *Kronecker product* (sometimes called *direct product*)

$$A \otimes B = (a_{ij}B). \tag{1–6–1}$$

If A is $m \times n$ and B is $l \times p$ then $A \otimes B$ is $(ml) \times (np)$, partitioned into mn blocks as in (1–6–1). If C is $n \times q$ and D is $p \times r$ it is easy to show using Theorem 1–5–1 that

Theorem 1–6–1

$$(A \otimes B)(C \otimes D) = AC \otimes BD.$$

Other elementary results involving the Kronecker product are given below.

Theorem 1–6–2

$$(A \otimes B)^* = A^* \otimes B^*.$$

Defining the *Kronecker power* by

$$A^{[2]} = A \otimes A$$

$$A^{[k+1]} = A \otimes A^{[k]}, \; k = 2, 3, \ldots$$

then

Theorem 1–6–3

$$A^{[k+l]} = A^{[k]} \otimes A^{[l]}$$

$$(AB)^{[k]} = A^{[k]}B^{[k]}.$$

The second result in Theorem 1–6–3 is interesting since generally

$$(AB)^k \neq A^k B^k.$$

A theorem which is of frequent application now follows.

Theorem 1–6–4
Let A be an $n \times n$ matrix. Then

(i) $AX = B$, where X is $n \times m$, may be written $(A \otimes I_m)x = b$, where x is the column nm-vector composed of the rows of X taken in order and b is formed in an identical fashion from B.
(ii) $XA = B$, where X is $m \times n$, may be written $(I_m \otimes A')x = b$.

Exercise 1–6–1
Show that Kronecker products are associative:

$$A \otimes B \otimes C = (A \otimes B) \otimes C = A \otimes (B \otimes C)$$

and distributive:

$$A \otimes (B + C) = A \otimes B + A \otimes C.$$

1–7 Determinants

The most important scalar function of a square matrix is the determinant. Before this can be defined some preliminary definitions are necessary.

A *permutation* of the integers 1 to n is an arrangement $\theta_1, \theta_2, \ldots, \theta_n$ in which each integer occurs exactly once. There are $n!$ distinct permutations. The interchange of two integers θ_i, θ_j in a permutation is called a *transposition.* It can be shown that any permutation can be obtained from $1, 2, \ldots, n$ by a succession of k transpositions, and the *sign* $\varepsilon(\theta_1, \theta_2, \ldots, \theta_n)$ of the permutation is defined to be $+1$ or -1 according as k is even or odd. An equivalent definition is

$$\varepsilon(\theta_1, \theta_2, \ldots, \theta_n) = \operatorname{sgn} \prod_{1 \leqslant i < j \leqslant n} (\theta_j - \theta_i)$$

where the function *signum* x is defined as

$$\operatorname{sgn} x = \begin{cases} +1 & (x > 0) \\ 0 & (x = 0) \\ -1 & (x < 0). \end{cases}$$

The *determinant* of the nth-order matrix $A = (a_{ij})$ is then defined as

$$\det A = \sum \varepsilon(\theta_1, \theta_2, \ldots, \theta_n) a_{1\theta_1} a_{2\theta_2} \ldots a_{n\theta_n},$$

the sum being taken over all the $n!$ permutations of $1, 2, \ldots, n$. The notation $|A|$ or

$$\begin{vmatrix} a_{11} & a_{12} & \cdots & a_{1n} \\ a_{21} & a_{22} & \cdots & a_{2n} \\ \cdot & \cdot & \cdots & \cdot \\ \cdot & \cdot & \cdots & \cdot \\ \cdot & \cdot & \cdots & \cdot \\ a_{n1} & a_{n2} & \cdots & a_{nn} \end{vmatrix}$$

is also used.

Some useful elementary properties of determinants are now given.

Theorem 1–7–1
Det $(A') = \det A$, $\det (A^*) = \overline{\det A}$.

Theorem 1–7–2
Det $(kA) = k^n \det A$ for any scalar k.

Theorem 1–7–3
Let B be the matrix obtained from A by the *elementary operations*:

(*a*) Interchange any two rows (or columns) of A. Then $\det B = -\det A$.
(*b*) Multiply any row (column) of A by a non-zero scalar k. Then $\det B = k \det A$.
(*c*) Add to any row (column) of A an arbitrary multiple of any other row (column). Then $\det B = \det A$.

Theorem 1–7–4
If A has any two rows (columns) identical or any row (column) zero then $\det A = 0$.

Theorem 1–7–5
Det $(AB) = \det A \det B$ for any two square matrices A, B of order n.

Theorem 1–7–6
If A is diagonal or triangular $\det A = a_{11}a_{22} \cdots a_{nn}$.

Theorem 1–7–7
If A, B are of orders m, n respectively then $\det (A \otimes B) = (\det A)^n(\det B)^m$.

A kth-order *minor* of A is the determinant of a kth-order submatrix of A and may be written in the notation of Section 1–5 as

$$D(i_1, i_2, \ldots, i_k \mid j_1, j_2, \ldots, j_k) = \det A(i_1, i_2, \ldots, i_k \mid j_1, j_2, \ldots, j_k). \quad (1\text{–}7\text{–}1)$$

A *principal minor* is a minor whose diagonal is part of the principal diagonal of A. The *cofactor* of the minor (1–7–1) is defined as

$$\tilde{D}(i_1, i_2, \ldots, i_k \mid j_1, j_2, \ldots, j_k) = (-1)^{i_1 + \cdots + i_k + j_1 + \cdots + j_k} D(i_{k+1}, \ldots, i_n \mid j_{k+1}, \ldots, j_n), \quad (1\text{–}7\text{–}2)$$

where $i_{k+1}, \ldots, i_n$ are the n–k numbers among $1, \ldots, n$ other than $i_1, \ldots, i_k$ and $j_{k+1}, \ldots, j_n$ are the n–k numbers among $1, \ldots, n$ other than $j_1, \ldots, j_k$. In particular when $k = n - 1$ it is common to write M_{ij} for the $(n-1) \times (n-1)$ determinant (1–7–1), which is then the determinant of the matrix obtained from A by deleting the ith row and jth column. Another case of importance is when $k = 1$, the expression (1–7–2) then giving

$$\tilde{D}(i \mid j) = (-1)^{i+j} M_{ij}, \quad (1\text{–}7\text{–}3)$$

which is written $$A_{ij} = (-1)^{i+j} M_{ij}, \tag{1–7–4}$$
being the cofactor of the element a_{ij}.

Theorem 1–7–8
Laplace's expansion theorem states that if $i_1, \ldots, i_k$ are integers such that $1 \leqslant k \leqslant n$ and $1 \leqslant i_1 < \cdots < i_k \leqslant n$, then

$$\det A = \sum_{1 \leqslant j_1 < \cdots < j_k \leqslant n} D(i_1, \ldots, i_k \mid j_1, \ldots, j_k) \tilde{D}(i_1, \ldots, i_k \mid j_1, \ldots, j_k). \tag{1–7–5}$$

That is, a determinant can be expanded in terms of k specified rows (numbered here $i_1, \ldots, i_k$), being the sum of the

$$\binom{n}{k} = \frac{n!}{k!(n-k)!}$$

products of each possible kth-order minor involving all the specified rows and its corresponding cofactor.

For example, let

$$M = \begin{bmatrix} A & 0 \\ B & C \end{bmatrix}$$

where each submatrix is $n \times n$, and expand $\det M$ by the first n rows. The only non-zero minor involving these rows is

$$D(1, 2, \ldots, n \mid 1, 2, \ldots, n) = \det A$$

and its cofactor is

$$\begin{aligned} \tilde{D}(1, 2, \ldots, n \mid 1, 2, \ldots, n) &= (-1)^{1+2+\cdots+n+1+2+\cdots+n} \\ &\qquad \times D(n+1, \ldots, 2n \mid n+1, \ldots, 2n) \\ &= (-1)^{n(n+1)} \det C. \end{aligned}$$

Thus, using (1–7–5), $\det M = \det A \det C$. This example is in fact a special case of more general formulae given later in Section 1–10.

A special case of Theorem 1–7–8 of great importance is when $k = 1$. Eqns (1–7–3), (1–7–4), and (1–7–5) then give:

Theorem 1–7–9
(Expansion by any row or column)

$$\sum_{r=1}^{n} a_{ir} A_{jr} = \delta_{ij} \det A \tag{1–7–6}$$

and $$\sum_{r=1}^{n} a_{ri} A_{rj} = \delta_{ij} \det A. \tag{1–7–7}$$

(δ_{ij} has been defined in Section 1–1.)

That is, the scalar product of any row (column) with the cofactors of the elements in that row (column) gives $\det A$ but the product of any row (column) with the cofactors of any other row (column) is zero.

Exercise 1–7–1

The *Vandermonde determinant* is defined by

$$D(a_1, a_2, \ldots, a_n) = \begin{vmatrix} 1 & a_1 & a_1^2 & \ldots & a_1^{n-1} \\ 1 & a_2 & a_2^2 & \cdots & a_2^{n-1} \\ \cdot & \cdot & \cdot & \cdots & \cdot \\ \cdot & \cdot & \cdot & \cdots & \cdot \\ \cdot & \cdot & \cdot & \cdots & \cdot \\ 1 & a_n & a_n^2 & \cdots & a_n^{n-1} \end{vmatrix}.$$

Subtract $a_1 \times$ column $(n - 1)$ from column n, then $a_1 \times$ column $(n - 2)$ from column $(n - 1), \ldots$, then $a_1 \times$ column 1 from column 2. (Use Theorem 1–7–3.) Expand the resulting determinant by the first row to show that

$$D(a_1, a_2, \ldots, a_n) = (a_2 - a_1)(a_3 - a_1) \ldots (a_n - a_1) D(a_2, a_3, \ldots, a_n).$$

Hence, prove by induction that

$$D(a_1, a_2, \ldots, a_n) = \prod_{1 \leqslant i < j \leqslant n} (a_j - a_i).$$

Exercise 1–7–2

Show that the determinant of a skew-symmetric matrix of odd order is zero, and that the determinant of a skew-hermitian matrix of odd order is pure imaginary.

1–8 Trace

Another useful scalar associated with a square matrix A is the *trace*, the sum of the elements on the principal diagonal of A, denoted by

$$\text{tr } A = \sum_{i=1}^{n} a_{ii}.$$

Some properties of trace are:

Theorem 1–8–1

For square matrices A, B, and scalars k_1, k_2,

$$\text{tr } (k_1 A + k_2 B) = k_1 \text{ tr } A + k_2 \text{ tr } B.$$

Theorem 1–8–2

For matrices A, B respectively $m \times n$ and $n \times m$,

$$\text{tr } (AB) = \text{tr } (BA).$$

Theorem 1–8–3

For an $m \times n$ matrix A,

$$\text{tr } (A\,A^*) = \text{tr } (A^*\,A) = \sum_{i=1}^{m} \sum_{j=1}^{n} | a_{ij} |^2.$$

The number $(\text{tr } (A\,A^*))^{1/2}$ is called the *euclidean norm* of A (see Section 2–9).

Theorem 1–8–4
For A and B square,

$$\text{tr}\,(A \otimes B) = (\text{tr}\,A)(\text{tr}\,B).$$

Exercise 1–8–1
Show that $\text{tr}\,(AB)^k = \text{tr}\,(BA)^k$, $k = 2, 3, \ldots$ (use Theorem 1–8–2).

1–9 Inverse

If A is a square matrix of order n and there exists a matrix B such that

$$AB = BA = I_n,$$

then A is said to be *non-singular* and B is called the *inverse* of A and written A^{-1}. If no such B exists, A is said to be *singular*.

Theorem 1–9–1
A matrix is non-singular if and only if $\det A \neq 0$.

The *adjoint* (or *adjugate*) of A is defined by

$$\text{adj}\,A = (A_{ji}),$$

that is, the transpose of the matrix of cofactors of the elements of A. It follows at once as a consequence of (1–7–6) and (1–7–7) that:

Theorem 1–9–2

$$A^{-1} = \text{adj}\,A/\det A. \tag{1–9–1}$$

Setting $B = A^{-1}$ in Theorem 1–3–1 immediately leads to:

Theorem 1–9–3

$$(A^*)^{-1} = (A^{-1})^*.$$

Some special forms of matrix previously defined have simple inverses. The inverse of

$$A = \text{diag}\,(a_{11}, a_{22}, \ldots, a_{nn})$$

is

$$A^{-1} = \text{diag}\,(1/a_{11}, 1/a_{22}, \ldots, 1/a_{nn})$$

provided all $a_{ii} \neq 0$. The inverse of a unitary matrix U is just U^*, and in particular the inverse of an orthogonal matrix T is T'.

It is sometimes worth while to invert a matrix by partitioning. Let

$$A = \begin{bmatrix} A_1 & A_2 \\ A_3 & A_4 \end{bmatrix},$$

where A_1, A_4 are non-singular square matrices of orders p, q. Then

$$A^{-1} = \begin{bmatrix} B_1 & B_2 \\ B_3 & B_4 \end{bmatrix},$$

where B_1, B_4 are also square matrices of orders p, q, and

$$B_1 = (A_1 - A_2 A_4^{-1} A_3)^{-1}$$
$$B_2 = -B_1 A_2 A_4^{-1}$$
$$B_3 = -A_4^{-1} A_3 B_1$$
$$B_4 = A_4^{-1} - A_4^{-1} A_3 B_2.$$

Thus only matrices of order p and q need to be inverted, rather than the original matrix of order $p + q$.

The inverse B of a non-singular lower (upper) triangular matrix A is also lower (upper) triangular and in either case is easily found algebraically. For example, when A is lower triangular, the rows of A^{-1} may be found in successive order from consideration of the identity $A \,.\, A^{-1} \equiv I_n$. In particular the principal diagonal of A^{-1} is

$$(1/a_{11}, 1/a_{22}, \ldots, 1/a_{nn}).$$

Exercise 1–9–1
Verify that if $A^{-1} = (b_{ij})$ then in the case when A is lower triangular,

$$b_{ij} = -(\sum_{k=j}^{i-1} a_{ik}b_{kj})/a_{ii} \quad (i > j).$$

It is convenient to give the following definition here: If A and B are square matrices and

$$B = S^{-1}AS, \tag{1–9–2}$$

where S is some square non-singular matrix, then A and B are said to be *similar*, and connected by a *similarity transformation* with transforming matrix S.

Theorem 1–9–4
Tr $(S^{-1}AS) =$ tr A, for S non-singular.

Exercise 1–9–2
Show that $(A \otimes B)^{-1} = A^{-1} \otimes B^{-1}$. (Hint: use Theorem 1–6–1.)

1–10 Partitioned determinants

The following theorems on the evaluation of

$$\Delta = \begin{vmatrix} A & B \\ C & D \end{vmatrix},$$

where A is $n \times n$, B is $n \times m$, C is $m \times n$, and D is $m \times m$ are often useful.

Theorem 1–10–1

$$\Delta = |A||D - CA^{-1}B| \quad \text{if} \quad |A| \neq 0$$
$$= |D||A - BD^{-1}C| \quad \text{if} \quad |D| \neq 0.$$

Theorem 1–10–2 (Schur's formulae)
When $m = n$

$$\begin{aligned}\Delta &= |AD - CB| \quad \text{if} \quad AC = CA \\ &= |DA - CB| \quad \text{if} \quad AB = BA \\ &= |AD - BC| \quad \text{if} \quad CD = DC \\ &= |DA - BC| \quad \text{if} \quad BD = DB.\end{aligned}$$

Theorem 1–10–3
When $m = 1$, so that B and C are vectors and D a scalar, Δ is said to be the result of *bordering A*. The first formula of Theorem 1–10–1 then reduces to

$$\Delta = d\,|A| - c\,Adj\,(A)b.$$

Exercise 1–10–1
For a given $n \times n$ matrix A with A_{ij} the cofactor of element a_{ij}, let

$$M = \begin{bmatrix} A_{11} & \cdots & A_{1m} \\ \cdot & \cdots & \cdot \\ \cdot & \cdots & \cdot \\ \cdot & \cdots & \cdot \\ A_{m1} & \cdots & A_{mm} \end{bmatrix}, \qquad N = \begin{bmatrix} A_{1,m+1} & \cdots & A_{m,m+1} \\ \cdot & \cdots & \cdot \\ \cdot & \cdots & \cdot \\ \cdot & \cdots & \cdot \\ A_{1n} & \cdots & A_{mn} \end{bmatrix},$$

and

$$P = \begin{bmatrix} M' & 0 \\ N & I_{n-m} \end{bmatrix}.$$

Show that det P = det M, and that

$$AP = \begin{bmatrix} |A|I_m & A_1 \\ 0 & A_2 \end{bmatrix},$$

where $A_1 = A(1, 2, \ldots, m \mid m + 1, m + 2, \ldots, n)$

$A_2 = A(m + 1, m + 2, \ldots, n \mid m + 1, m + 2, \ldots, n)$

(use Theorem 1–7–9). Using Theorem 1–10–1 show that det $(AP) = |A|^m |A_2|$ and hence that $|M| = |A|^{m-1} |A_2|$. This result is known as *Jacobi's Theorem.*

Exercise 1–10–2
By taking $m = 2$ in Jacobi's Theorem prove by induction on n that if A is skew-symmetric and of even order its determinant is a perfect square (use Exercise 1–7–2). In fact the determinant of a skew-symmetric matrix of even order can be shown to be the square of a polynomial function of its elements, the function being called the *Pfaffian* [Ferrar, (1941)].

1–11 Linear dependence and rank

Consider the set of n-vectors

$$a_i = (a_{i1}, a_{i2}, \ldots, a_{in}), \qquad i = 1, 2, \ldots, m.$$

If there exist scalars $c_1, c_2, \ldots, c_m$, not all zero, such that $\sum_{i=1}^{m} c_i a_i = 0$, then $a_1, \ldots, a_m$ are said to be *linearly dependent*. Otherwise $a_1, \ldots, a_m$ are *linearly independent*.

The rank $R(A)$ of an $m \times n$ matrix A is the order of the largest non-singular submatrix of A. A non-vanishing minor of A of maximum order is termed a *critical* minor. Clearly $R(A) = 0$ if and only if $A = 0$.

An essential result is

Theorem 1–11–1
The rank of A is equal to the maximum value of r for which there are r linearly independent rows or columns.

In particular:

Theorem 1–11–2
A square matrix is singular if and only if its rows (or columns) are linearly dependent.

A simple test for determining whether a given set of m n-vectors is linearly dependent or not is to form a matrix A from these vectors. Then:

Theorem 1–11–3
A necessary and sufficient condition for the vectors to be linearly dependent is $R(A) < m$.

Some other useful theorems concerning the rank of a matrix are now given.

Theorem 1–11–4
If A is $m \times n$ and B is $n \times p$, then

$$R(A) + R(B) - n \leqslant R(AB) \leqslant \min(R(A), R(B)).$$

Theorem 1–11–5
If A, B are both $m \times n$, then

$$R(A + B) \leqslant R(A) + R(B).$$

Theorem 1–11–6
The rank of a matrix is unaltered if it is pre- or post-multiplied by a non-singular square matrix.

Theorem 1–11–7
For any two matrices

$$R(A \otimes B) = R(A)\, R(B).$$

Exercise 1–11–1
For rectangular matrices A_i, if $A = \sum_{i=1}^{k} \oplus A_i$, show that $R(A) = \sum_{i=1}^{k} R(A_i)$.

Exercise 1–11–2
Let A be a square matrix of order n and b a vector such that $A^k b \neq 0$, $A^{k+1} b = 0$ for some integer $k < n$. Show that the vectors $b, Ab, \ldots, A^k b$ are linearly independent. (Hint: assume $\sum c_i A^i b = 0$ and show this implies $c_0 = 0$, $c_1 = 0, \ldots, c_k = 0$.)

1–12 Linear equations

A system of m linear equations in n unknowns

$$\sum_{j=1}^{n} a_{ij} x_j = b_i, \qquad i = 1, 2, \ldots, m$$

can be written in vector-matrix form as

$$Ax = b, \tag{1–12–1}$$

where $A = (a_{ij})$ is the matrix of coefficients, x is the column n-vector with components x_j, b the column m-vector with components b_i. It is useful to define the *augmented matrix* $B = (A, b)$. The system is called *homogeneous* if $b = 0$. A system of equations is called *consistent* if it possesses at least one solution (which may include $x = 0$), otherwise it is called *inconsistent.*

The basic result is:

Theorem 1–12–1
The system (1–12–1) possesses

(*a*) an infinite number of solutions if and only if $R(A) = R(B) < n$;
(*b*) a unique solution if and only if $R(A) = R(B) = n$;
(*c*) no solution if and only if $R(A) < R(B)$.

In the particular case when A is square and non-singular the unique solution of (1–12–1) is $x = A^{-1}b$. This can also be written as

$$x_i = |A^{(i)}| / |A|, \qquad i = 1, 2, \ldots, n, \tag{1–12–2}$$

where $A^{(i)}$ is the matrix obtained by replacing the ith column of A with the vector b. The equation (1–12–2) is known as *Cramer's rule.*

Theorem 1–12–2
The homogeneous system

$$Ax = 0 \tag{1–12–3}$$

possesses a non-zero solution if and only if $R(A) < n$.

An immediate consequence of Theorem 1–12–2 is that if A is square the system (1–12–3) has a non-zero solution if and only if A is singular; further, if $m < n$ then (1–12–3) always has a non-zero solution.

Generally, to solve (1–12–1)

(i) find a particular solution $x^{(0)}$ to (1–12–1);

(ii) find the $k = n - R(A)$ linearly independent solutions

$$x^{(1)}, x^{(2)}, \ldots, x^{(k)} \text{ of (1–12–3);}$$

(iii) the general solution of (1–12–1) is then

$$x = \sum_{i=1}^{k} c_i x^{(i)} + x^{(0)}$$

where the c_i are arbitrary scalars.

1–13 Equivalence and congruence

Elementary operations on a matrix have already been defined in Theorem 1–7–3, and it follows at once from that theorem that the rank of a matrix is unaltered by elementary transformations. An *elementary matrix* is any matrix derived from a unit matrix by a single elementary operation.

Theorem 1–13–1

Every elementary matrix is non-singular, and the inverse of an elementary matrix is again an elementary matrix.

To bring about an elementary transformation on a given matrix A, the same transformation is first applied to the unit matrix of suitable order, and the resulting matrix then pre-multiplies A for row operations and post-multiplies for columns.

A matrix A is said to be *equivalent* to B (written $A \sim B$) if it is possible to pass from A to B by a chain of elementary operations. The three important properties of equivalence follow at once:

(i) Reflexive, i.e., $A \sim A$;

(ii) Symmetric, i.e., $A \sim B$ implies $B \sim A$;

(iii) Transitive, i.e., $A \sim B$ and $B \sim C$ imply $A \sim C$.

Theorem 1–13–2

Two $m \times n$ matrices A and B are equivalent if and only if

either (i)* there exist non-singular matrices L, M of orders m, n respectively such that

$$LAM = B \qquad (1–13–1)$$

or (ii) $$R(A) = R(B).$$

When A and B satisfy an equation of the form (1–13–1) they are said to be related by an equivalence transformation.

* This is often used as the definition of equivalence.

The next theorem follows easily from Theorem 1–13–2.

Theorem 1–13–3
If A is an $m \times n$ matrix of rank r there exist non-singular matrices L, M of orders m, n respectively such that

$$LAM = \begin{bmatrix} I_r & 0 \\ 0 & 0 \end{bmatrix}, \tag{1–13–2}$$

where the zero matrices are of suitable order.

A particular case of Theorem 1–13–3 is:

Theorem 1–13–4
If A is square and non-singular then it can be expressed as a product of elementary matrices.

Comparison of (1–13–1) and (1–9–2) immediately leads to:

Theorem 1–13–5
Similar matrices are equivalent.

An important special class of equivalence transformation when A and B are square and of the same order is called a *congruence* transformation, defined by

$$B = L'AL, \tag{1–13–3}$$

where L is non-singular; A and B are said to be congruent.

A result which will be referred to later is:

Theorem 1–13–6
A matrix is congruent to a diagonal matrix if and only if it is symmetric.

Exercise 1–13–1
A square matrix A is called *idempotent* if $A^2 = A$. Use Theorem 1–13–3 to show that an idempotent matrix of rank r has trace equal to r. (Hint: use Theorem 1–8–2.)

Discussion and further reading

Statements of elementary but important facts in matrix algebra have been given. All are very well known and may be found, together with proofs, in standard texts such as those by Gantmacher (1959), Mirsky (1963), Finkbeiner (1966), and Bellman (1960). The intention here was not to duplicate such volumes but to present basic information on matrices in a convenient and concise form.

Older works which may also be consulted with profit include Ferrar (1941, 1951), Frazer, Duncan, and Collar (1938), Wedderburn (1934), and

Bôcher (1907). A very comprehensive account containing statements of most of the results of this and some later chapters, together with some proofs and many references, is given by Marcus and Minc (1964). See also Rosenbrock and Storey (1970), MacDuffee (1956), and Wilkinson (1965). Another useful compendium which also contains material on polynomials and on abstract algebra is that by Mishina and Proskuryakov (1965).

Numerical problems of linear algebra, dealt with only in passing in this book are considered by Fox (1964) and more fully by Faddeev and Faddeeva (1963).

References

BELLMAN, R. (1960), *Introduction to matrix analysis*, McGraw-Hill, New York.

BÔCHER, M. (1907, reprinted 1964), *Introduction to higher algebra*, Dover, New York.

FADDEEV, D. K. and V. N. FADDEEVA (1963), *Computational methods of linear algebra*, W. H. Freeman, San Francisco.

FERRAR, W. L. (1941), *Algebra*, Oxford University Press, London.

FERRAR, W. L. (1951), *Finite matrices*, Oxford University Press, London.

FINKBEINER, D. T. (1966), *Introduction to matrices and linear transformations*, 2nd edn., W. H. Freeman, San Francisco.

FOX, L. (1964), *An introduction to numerical linear algebra*, Oxford University Press, London.

FRAZER, R. A., W. J. DUNCAN and A. R. COLLAR (1938), *Elementary matrices*, Cambridge University Press, London.

GANTMACHER, F. R. (1959), *The theory of matrices*, Vol. I, Chelsea Publishing Co., New York.

MACDUFFEE, C. C. (1956), *The theory of matrices*, Chelsea Publishing Co., New York.

MARCUS, M. and H. MINC (1964), *A survey of matrix theory and matrix inequalities*, Allyn and Bacon, Boston.

MIRSKY, L. (1963), *An introduction to linear algebra*, Oxford University Press, London.

MISHINA, A. P. and I. V. PROSKURYAKOV 196(5), *Higher algebra*, Pergamon Press, Oxford.

ROSENBROCK, H. H. and C. STOREY (1970), *Mathematics of dynamical systems*, Nelson, London.

WEDDERBURN, J. H. M. (1934, reprinted 1964), *Lectures on matrices*, Dover, New York.

WILKINSON, J. H. (1965), *The algebraic eigenvalue problem*, Oxford University Press, London.

2

Matrix algebra : further theory

2–1 Decomposable matrices

A matrix for which $a_{i\theta_i} = 1$ ($i = 1, 2, \ldots, n$), θ_i being a permutation of the integers 1 to n, and $a_{ij} = 0$ otherwise, is called a *permutation matrix*. In other words, such a matrix has exactly one single unit element in each row and column, all other entries being zero.

Theorem 2–1–1
If P is a permutation matrix so is P' and $P'P = I$.

A square matrix A is said to be *decomposable* (sometimes *reducible*) if there exists a permutation matrix P such that

$$PAP' = \begin{bmatrix} B & 0 \\ C & D \end{bmatrix} \tag{2–1–1}$$

where B and D are square. Otherwise A is *indecomposable*. Notice that $P'AP$ is the matrix A with row i replaced by row θ_i, column i replaced by column θ_i ($i = 1, 2, \ldots, n$).

Theorem 2–1–2
An indecomposable matrix has at least one non-zero off diagonal element in each row and column.

Another useful result is

Theorem 2–1–3
If H is a hermitian matrix with $R(H) \geqslant k$ then there exists a permutation matrix P such that $P'HP$ has a non-singular matrix of order k in the upper left corner.

The definition of decomposability is closely related to the solution of a set of linear equations

$$Ax = b, \tag{2–1–2}$$

where A is the matrix in (2–1–1). Setting $x = P'y$, $b = P'c$ and using Theorem 2–1–1 transforms (2–1–2) into

$$\begin{bmatrix} B & 0 \\ C & D \end{bmatrix} \begin{bmatrix} y_1 \\ y_2 \end{bmatrix} = \begin{bmatrix} c_1 \\ c_2 \end{bmatrix} \tag{2–1–3}$$

where y_1, y_2, c_1, c_2 are suitably dimensioned column vectors. Clearly (2–1–3) has solution $y_1 = B^{-1}c_1$, $y_2 = D^{-1}(c_2 - Cy_1)$ provided B, D are non-singular, so that inversion of A can be reduced to inversion of the smaller order matrices B and D.

Exercise 2–1–1

Let the nth-order matrices $A = (a_{ij})$, $B = (b_{ij})$ be such that $a_{ij} \geqslant 0$, $b_{ij} > 0$, all i, j. (A is called *non-negative* and B *positive*.) Show that if A is indecomposable then AB is also positive.

2–2 The characteristic equation

For an $n \times n$ matrix A over **C**, the *characteristic polynomial* of A is

$$\begin{aligned} p(\lambda) &= \det(\lambda I_n - A) \\ &= \lambda^n + p_1\lambda^{n-1} + \cdots + p_n. \end{aligned} \tag{2–2–1}$$

The matrix $\lambda I_n - A$ is called the *characteristic* matrix of A and the equation

$$p(\lambda) = 0 \tag{2–2–2}$$

the *characteristic equation* of A. It is easily shown that:

Theorem 2–2–1

$$p_1 = -\mathrm{tr}\,(A), \quad p_n = (-1)^n \det A$$

and generally p_k is $(-1)^k$ times the sum of all k-rowed principal minors of A.

Much more convenient expressions for the p_i may be obtained in terms of the inverse of the characteristic matrix. It follows from (1–9–1) that

$$(\lambda I_n - A)^{-1} = B(\lambda)/p(\lambda) \tag{2–2–3}$$

where

$$\begin{aligned} B(\lambda) &= \mathrm{Adj}\,(\lambda I_n - A) \\ &= \lambda^{n-1}I_n + \lambda^{n-2}B_1 + \lambda^{n-3}B_2 + \cdots + B_{n-1}. \end{aligned} \tag{2–2–4}$$

Then:

Theorem 2–2–2 (Leverrier's Algorithm)

$p_1, \ldots, p_n$ and $B_1, \ldots, B_{n-1}$ are determined by

$$B_1 = A + p_1 I_n,\ B_k = AB_{k-1} + p_k I_n,\ k = 2, 3, \ldots, n-1 \tag{2–2–5}$$

and

$$p_1 = -\mathrm{tr}\,(A), \quad p_k = -\frac{1}{k}\mathrm{tr}\,(AB_{k-1}),\ k = 2, 3, \ldots, n. \tag{2–2–6}$$

Thus $p_1, p_2, \ldots, p_n$ may be calculated in succession; also, a useful check for computational purposes is provided by the fact that if k is set equal to n in (2–2–5) then B_n should be the nth-order zero matrix if no rounding errors are present.

Theorem 2–2–3

Similar matrices have the same characteristic equation.

It then easily follows that:

Theorem 2–2–4
For arbitrary nth-order matrices A and B, the characteristic polynomials of AB and BA are the same.

If $f(t) = f_0 + f_1 t + \cdots + f_m t^m$ then $f(A) = f_0 I_n + f_1 A + \cdots + f_m A^m$ is a *polynomial in the square matrix A*. A polynomial f is said to *annihilate A* if $f(A) = 0$. The unique monic* polynomial of least degree which annihilates A is called the *minimum polynomial* of A.

Theorem 2–2–5
Similar matrices have the same minimum polynomial.

Theorem 2–2–6 (Cayley-Hamilton Theorem)
Every square matrix A satisfies its own characteristic equation, i.e.,

$$A^n + p_1 A^{n-1} + \ldots + p_n I_n = 0, \tag{2–2–7}$$

which can be written $p(A) = 0$.

Notice that (2–2–7) gives an expression for A^n in terms of $I_n, A, A^2, \ldots, A^{n-1}$. By multiplying both sides of (2–2–7) by A and using the expression for A^n, an expression can be obtained for A^{n+1} also in terms of $I_n, A, A^2, \ldots, A^{n-1}$. By repeating this process it can be seen that any power of A greater than $n - 1$ can be expressed as a linear combination of $I_n, A, \ldots, A^{n-1}$, so that any polynomial in A can be reduced to a polynomial of degree not exceeding $n - 1$. Similarly, by multiplying both sides of (2–2–7) by A^{-1}, the inverse of A can also be expressed as a linear combination of the first $n - 1$ powers.

An immediate consequence of Theorem 2–2–6 is

Theorem 2–2–7
The minimum polynomial of A is a factor of the characteristic polynomial of A.

A matrix for which the minimum and characteristic polynomials are identical is called *non-derogatory*; otherwise *derogatory*.

Since the ground field is taken to be $\mathbf{C}$ it follows from a fundamental result of the algebra of polynomials that $p(\lambda)$ can be written as a product of linear factors, some possibly occurring more than once, i.e.,

$$p(\lambda) = (\lambda - \lambda_1)^{i_1}(\lambda - \lambda_2)^{i_2} \cdots (\lambda - \lambda_r)^{i_r}, \; i_k \geqslant 1.$$

Then

Theorem 2–2–8
The distinct linear factors of the minimum polynomial coincide with those of the characteristic polynomial.

* A monic polynomial has unity as the coefficient of its highest degree term.

Exercise 2–2–1

A square matrix A is called *nilpotent* to index r if $A^r = 0$, $A^{r-1} \neq 0$ for some positive integer r. Show that the minimum polynomial of A is λ^r and that

$$(I - A)^{-1} = I + A + A^2 + \cdots + A^{r-1}.$$

Exercise 2–2–2

A *cyclic* arrangement of the ordered set $\alpha_1, \alpha_2, \ldots, \alpha_k$ is any arrangement of the form

$$\alpha_i, \alpha_{i+1}, \ldots, \alpha_k, \alpha_1, \alpha_2, \ldots, \alpha_{i-1}.$$

If $A_1, A_2, \ldots, A_k$ is an ordered set of matrices deduce from Theorem 2–2–4 that all cyclic products have the same characteristic polynomial. Using Theorem 2–2–1 write down a corresponding generalization of Theorem 1–8–2.

Exercise 2–2–3

Let $g(A)$ be a non-singular matrix polynomial in the matrix A, and let $m(\lambda) = \lambda^r + m_1\lambda^{r-1} + \cdots + m_{r-1}\lambda + m_r$ be the minimum polynomial of $g(A)$. Deduce that $m_r \neq 0$. Show that $(g(A))^{-1}$ is also equal to a polynomial in A, and hence that $f(A)(g(A))^{-1}$ or $(g(A))^{-1}f(A)$, for any polynomial f, is also a polynomial in A (and thus may be written symbolically as $f(A)/g(A)$).

2–3 Characteristic roots

The n roots of the characteristic equation (2–2–1) are called the *characteristic roots* (or *eigenvalues, latent roots, proper values*, or *secular values*) of A and are denoted by $\lambda_i(A)$. Since the ground field is **C**.

$$p(\lambda) \equiv (\lambda - \lambda_1)(\lambda - \lambda_2) \cdots (\lambda - \lambda_n), \tag{2–3–1}$$

where some of the factors may occur more than once, and it then follows immediately from (2–3–1) and Theorem (2–2–1) that

Theorem 2–3–1

$$-p_1 = \text{tr}\,(A) = \sum_{i=1}^{n} \lambda_i(A) \tag{2–3–2}$$

$$(-1)^n p_n = \det A = \prod_{i=1}^{n} \lambda_i(A). \tag{2–3–3}$$

Similar expressions can be obtained for $p_2, \ldots, p_{n-1}$.

Some properties of characteristic roots are now given.

Theorem 2–3–2

A is singular if and only if at least one $\lambda_i(A)$ is zero.

This follows at once from (2–3–3).

Theorem 2–3–3

If A is diagonal or triangular its characteristic roots are $a_{11}, a_{22}, \ldots, a_{nn}$.

This follows from Theorem 1–7–6.

Theorem 2–3–4

The characteristic roots of a hermitian matrix (and in particular of a real symmetric matrix) are all real.

Theorem 2–3–5

The characteristic roots of a skew-hermitian matrix are purely imaginary or zero.

Theorem 2–3–6

The characteristic roots of a unitary matrix lie on the unit circle in the complex plane (i.e., have unit modulus).

Theorem 2–3–7

A nilpotent matrix (see Exercise 2–2–1) has all characteristic roots zero.

Theorem 2–3–8

An idempotent matrix (see Exercise 1–13–1) has all characteristic roots either 0 or 1.

Theorem 2–3–9

If A is non-singular then $\lambda_i(A^m) = (\lambda_i(A))^m$ for any integer $m \neq 0$. (If m is positive the result still holds for A singular.)

Theorem 2–3–10

If $g(t) = a_0 + a_1 t + \cdots + a_m t^m$ then the characteristic roots of

$$g(A) = a_0 I_n + \cdots + a_m A^m \text{ are } g(\lambda_i(A)).$$

Theorem 2–3–11

Similar matrices have the same characteristic roots.

Theorem 2–3–12

$$\lambda_i(A^*) = \overline{\lambda_i(A)}$$

so

$$\lambda_i(A') = \lambda_i(A) \text{ and } \lambda_i(\overline{A}) = \overline{\lambda_i(A)}.$$

Theorem 2–3–13

The characteristic roots of $\sum_{j=1}^{n} \oplus A_j$ are just the set of $\lambda_i(A_j)$ taken together.

Theorem 2–3–14

If A is of order n, B of order m then the characteristic roots of $A \otimes B$ are the mn numbers $\lambda_i(A)\lambda_j(B)$, $i = 1, \ldots, n; j = 1, \ldots, m$.

Theorem 2–3–15

If A is of order n, B of order m then the characteristic roots of $A \otimes I_m + I_n \otimes B$ are the mn numbers $\lambda_i(A) + \lambda_j(B)$, $i = 1, \ldots, n; j = 1, \ldots, m$.

Apart from matrices of special form (e.g., Theorems 2–3–4, 5, 6 above), the roots of a general matrix can lie anywhere in the complex plane. However, there is a considerable body of results on the localization of roots of a general matrix. The object is to provide information about the location of characteristic roots in the complex plane without actual calculation of the roots. Chapter 3 of Marcus and Minc (1964) is devoted entirely to this topic and Householder gives an up-to-date survey in Schneider (1964); both accounts give many references. Only what is perhaps the best-known result, a theorem due to Gershgorin, is given here.

Theorem 2–3–16

Every root of a complex matrix A lies in at least one of the n disks with centres a_{ii}, radii $\rho_i = \sum_{\substack{j=1 \\ j \neq i}}^{n} | a_{ij} |$.

A sometimes sharper set of bounds can be obtained by applying this theorem to A and A' simultaneously.

A consequence of Theorem 2–3–16 is that if A is *dominated* by its diagonal elements, i.e., $| a_{ii} | > \sum_{\substack{j=1 \\ j \neq i}}^{n} | a_{ij} |$, $i = 1, 2, \ldots, n$ then each $\lambda_i(A)$ satisfies at least one of the equations

$$| \lambda_i(A) - a_{ii} | < | a_{ii} |, i = 1, \ldots, n.$$

This implies that no $\lambda_i(A)$ is zero, so that by Theorem 2–3–2 A is non-singular.

For example,

$$A = \begin{bmatrix} -3 & 1 & 1 \\ 2 & -5 & -2 \\ -1 & -2 & -4 \end{bmatrix}$$

is dominated by its diagonal elements and by Theorem 2–3–16 each root λ lies in at least one of the disks $| \lambda + 3 | \leqslant 2$, $| \lambda + 5 | \leqslant 4$, $| \lambda + 4 | \leqslant 3$. Thus all three roots of A lie to the left of the imaginary axis – that is, have negative real parts (such a matrix is called *stable*).

Exercise 2–3–1

If the characteristic roots of an $(n + 1) \times (n + 1)$ matrix A are zero and the nth roots of unity, show that

$$2I/(2I - A) = I + \frac{1}{2^{n-1}}(2^{n-1}A + 2^{n-2}A^2 + \cdots + 2A^{n-1} + A^n)$$

(see Exercise 2–2–3).

Exercise 2–3–2

If $\lambda_i (i = 1, 2, \ldots, n)$ are the characteristic roots of a matrix A prove that

$$\sum_{i=1}^{n} \lambda_i^2 = (\sum_{i=1}^{n} \lambda_i)^2 - 2\sum_{i<j} \lambda_i \lambda_j$$

and hence that $\sum_{i<j} \lambda_i \lambda_j = 2[(\text{tr } A)^2 - \text{tr }(A^2)]$.

(Hint: use Theorems 2–3–1, 2–3–9.)

Exercise 2–3–3

Use Theorem 2–3–16 to show that if a complex matrix $A = (a_{ij})$ satisfies $\text{Re}(a_{ii}) \leqslant -\rho_i$ then the real parts of all the characteristic roots of A are non-positive (A is called *semi-stable*).

2–4 Characteristic vectors

Any non-zero column n-vector x which satisfies the equation

$$(\lambda_i(A)I_n - A)x = 0$$

is called a *characteristic vector* (or *eigenvector*, *latent vector*, or *proper vector*). of A corresponding to the characteristic root λ_i. Some important properties of characteristic vectors are as follows:

Theorem 2–4–1

If $u_1, u_2, \ldots, u_k$ are characteristic vectors corresponding to distinct characteristic roots $\lambda_1, \lambda_2, \ldots, \lambda_k$ of $A (k \leqslant n)$ then $u_1, u_2, \ldots, u_k$ are linearly independent.

Theorem 2–4–2

If $k = n$ then the matrix $U = (u_1, u_2, \ldots, u_n)$ is such that $U^{-1}AU = \text{diag}(\lambda_1, \lambda_2, \ldots, \lambda_n)$ and if $U^{-1} = (\mathrm{v}_1, \mathrm{v}_2, \ldots, \mathrm{v}_n)'$ then $\mathrm{v}_1, \mathrm{v}_2, \ldots, \mathrm{v}_n$ are characteristic vectors of A' (corresponding to $\lambda_1, \lambda_2, \ldots, \lambda_n$).

If a characteristic root of A occurs k times it is said to have *multiplicity k*. There exist at most k_i linearly independent characteristic vectors associated with each root λ_i of multiplicity k_i. More precisely,

Theorem 2–4–3

$$R(\lambda_i I_n - A) \geqslant n - k_i. \qquad (2\text{–}4\text{–}1)$$

When equality occurs in (2–4–1), so that there are exactly k_i linearly independent characteristic vectors associated with λ_i, this root is termed *regular*. If $k_i = 1$, λ_i is said to be a *simple* root.

Theorem 2–4–4
A matrix A is similar to a diagonal matrix if and only if all its characteristic roots are regular; that is, if and only if A has n linearly independent characteristic vectors.

A matrix with fewer than n linearly independent characteristic vectors is called *defective*. Normal matrices are special cases of non-defective matrices and in fact a stronger version of Theorem 2–4–4 then applies:

Theorem 2–4–5
A matrix is unitarily similar to a diagonal matrix (i.e., the transforming matrix is unitary) if and only if it is normal.

A similar result, which is not however a special case of Theorem 2–4–5 holds for real symmetric matrices:

Theorem 2–4–6
Every real symmetric matrix is orthogonally similar to a diagonal matrix.

When A is defective (i.e., has multiple roots and is not similar to a diagonal matrix) then it is necessary to introduce the $k \times k$ *Jordan block matrix*

$$J_k(\lambda) = \begin{bmatrix} \lambda & 1 & 0 & \cdots & 0 & 0 \\ 0 & \lambda & 1 & \cdots & 0 & 0 \\ \cdot & \cdot & \cdot & \cdots & \cdot & \cdot \\ \cdot & \cdot & \cdot & \cdots & \cdot & \cdot \\ \cdot & \cdot & \cdot & \cdots & \cdot & \cdot \\ 0 & 0 & 0 & \cdots & \lambda & 1 \\ 0 & 0 & 0 & \cdots & 0 & \lambda \end{bmatrix}. \qquad (2\text{–}4\text{–}2)$$

It follows at once from Theorem 2–3–3 that $J_k(\lambda)$ has characteristic root λ of multiplicity k.

The basic result is:

Theorem 2–4–7
Every matrix A is similar to

$$J_{m_1}(\lambda_1) \oplus J_{m_2}(\lambda_1) \oplus \cdots \oplus J_{m_s}(\lambda_1) \oplus J_{n_1}(\lambda_2) \oplus \cdots \oplus J_{q_u}(\lambda_n) \qquad (2\text{–}4\text{–}3)$$

where the sum of the orders of the Jordan submatrices associated with λ_i is equal to the multiplicity k_i, i.e.,

$$m_1 + m_2 + \cdots + m_s = k_1, \text{ etc.}$$

The representation (2–4–3) is called the *Jordan normal* or *canonical* form of A and is often written

$$\operatorname{diag}(J_{m_1}(\lambda_1), \ldots, J_{q_u}(\lambda_n)).$$

The Jordan form is unique apart from the ordering of the blocks. The total number of linearly independent characteristic vectors of A is equal to the number of Jordan blocks in the Jordan form (2–4–3). Thus there are $s(\leqslant k_1)$ linearly independent characteristic vectors associated with the root λ_1 and $s = k_1$ only if each $m_i = 1$, etc.

For example, suppose a matrix A of order five has Jordan form

$$\begin{bmatrix} J_2(\lambda_1) & & \\ & J_2(\lambda_1) & \\ & & J_1(\lambda_2) \end{bmatrix}$$

and a second matrix B also of order five has Jordan form

$$\begin{bmatrix} J_3(\lambda_1) & & \\ & J_1(\lambda_1) & \\ & & J_1(\lambda_2) \end{bmatrix}.$$

Then A and B are not similar although both have a root λ_1 repeated four times and a root λ_2 occurring once. A and B both have three linearly independent characteristic vectors.

A matrix is derogatory (see Section 2–2) if there is more than one Jordan block (and therefore more than one characteristic vector) associated with λ_i for some i. Thus a non-derogatory matrix has only one Jordan block (and therefore only one characteristic vector) associated with each distinct λ_i.

The Jordan form is discussed more fully in Chapter 3.

Exercise 2–4–1

If $A^2 = I$ show that the characteristic roots of A are ± 1 and that if x is a non-zero column vector which is not a characteristic vector of A, then $x + Ax$ and $x - Ax$ are linearly independent characteristic vectors of A.

2–5 Polynomial matrices

Consider now an $m \times n$ matrix $A(\lambda)$ whose elements are polynomials in a single indeterminate λ with coefficients in **C**. If the highest degree of the elements of A is N then

$$A(\lambda) = A_0 + \lambda A_1 + \cdots + \lambda^N A_N, \qquad (2\text{–}5\text{–}1)$$

where the A_i are constant $m \times n$ matrices over **C**. Such a matrix $A(\lambda)$ is termed a *polynomial matrix* (or *lambda matrix*) of *degree* $\delta(A) = N$ over the polynomial domain $\mathbf{C}[\lambda]$. Clearly the characteristic matrix of Section 2–2 is an example of such a matrix. Addition, multiplication, etc. of polynomial matrices (if possible) are defined in obvious ways. For example,

$$A + B = \sum_i \lambda^i (A_i + B_i),$$

$$AB = \sum_i \lambda^i C_i \text{ where } C_i = \sum_{j+k=i} A_j B_k.$$

Division can be dealt with as follows:

Theorem 2–5–1
If A, B are square and of the same order and $|A_N| \neq 0$, then there exists one and only one pair of polynomial matrices Q_1, R_1 such that

$$B = Q_1 A + R_1$$

with $\delta(R_1) < \delta(A)$, and one and only one pair Q_2, R_2 such that

$$B = AQ_2 + R_2$$

with $\delta(R_2) < \delta(A)$.

A square polynomial matrix of order n is called *regular* if and only if $|A_N| \neq 0$, so that det $A(\lambda)$ is of maximum possible degree, namely nN. This usage of the term 'regular' is the more common one (e.g., Gantmacher (1959), Lancaster (1966)) but other authors (e.g., Hodge and Pedoe (1953)) unfortunately attach a quite different meaning to the word, calling A regular if its inverse is also a polynomial matrix. This implies, by virtue of (1–9–1) that the determinant of A is independent of λ, so that det A = det A_0. Thus the two definitions are mutually exclusive; it would seem preferable to call A *invertible* in the second situation. When A is invertible it follows from (1–9–2) that $\delta(A^{-1}) \leqslant N(n-1)$.

Generalizing the definition of Theorem 1–7–3, define the *elementary operations* on a rectangular polynomial matrix:

(i) Interchange any two rows (columns).
(ii) Multiply any row (column) by a non-zero scalar constant.
(iii) Add to any row (column) any other row (column) multiplied by an arbitrary polynomial $a(\lambda)$.

With this generalized definition, Theorem 1–13–1 still applies. As in the case of matrices over **C**(1–13) two $m \times n$ polynomial matrices $A(\lambda)$ and $B(\lambda)$ are said to be equivalent over $\mathbf{C}[\lambda]$ if and only if B can be obtained from A by a sequence of elementary operations. That is, if and only if

$$B(\lambda) = P(\lambda)\, A(\lambda)\, Q(\lambda), \tag{2–5–2}$$

where $P(\lambda)$ and $Q(\lambda)$ are invertible polynomial matrices. Notice that (2–5–2) then implies

$$A(\lambda) = P^{-1}(\lambda)\, B(\lambda)\, Q^{-1}(\lambda).$$

Another aspect of equivalence is now detailed. Let $A(\lambda)$ be an $m \times n$ polynomial matrix of rank r, and denote by $d_j(\lambda)$ the greatest common divisor of all jth-order minors of $A(\lambda)$, $j = 1, 2, \ldots, r$. The $d_j(\lambda)$ are polynomials called the *determinantal divisors* of $A(\lambda)$, and it can be assumed without loss of generality that each d_j is monic.

Theorem 2–5–2
d_j is divisible by $d_{j-1}, j = 1, 2, \ldots, r$ with d_0 taken to be unity.

The quotients

$$i_j(\lambda) = d_j/d_{j-1}, \quad j = 1, 2, \ldots, r \tag{2–5–3}$$

are also monic and are called the *invariant polynomials* of $A(\lambda)$.

This term arises since:

Theorem 2–5–3
Two matrices $A(\lambda)$, $B(\lambda)$ are equivalent if and only if they have the same invariant polynomials.

More generally, a most important fundamental result is:

Theorem 2–5–4
A matrix $A(\lambda)$ of rank r is equivalent to

$$\operatorname{diag}\,(i_1(\lambda), i_2(\lambda), \ldots, i_r(\lambda), 0, \ldots, 0), \tag{2–5–4}$$

called the *Smith normal* (or *canonical*) form of A. Moreover, $i_j(\lambda)$ divides $i_{j+1}(\lambda)$ for all j.

The basic idea behind the proof of Theorem 2–5–4 is contained in the following very simple example:

$$A = \begin{bmatrix} 2\lambda^2 - 6\lambda + 1 & 2\lambda \\ \lambda^2 - 3\lambda & \lambda \end{bmatrix}$$

$$\sim \begin{bmatrix} 2\lambda & 2\lambda^2 - 6\lambda + 1 \\ \lambda & \lambda^2 - 3\lambda \end{bmatrix} \qquad C1 \leftrightarrow C2$$

$$\sim \begin{bmatrix} 2\lambda & 1 \\ \lambda & 0 \end{bmatrix} \qquad C2 - (\lambda \quad 3)C1$$

$$\sim \begin{bmatrix} 1 & 2\lambda \\ 0 & \lambda \end{bmatrix} \qquad C1 \leftrightarrow C2$$

$$\sim \begin{bmatrix} 1 & 0 \\ 0 & \lambda \end{bmatrix}. \qquad C2 - (2\lambda)C1$$

The elementary operations which have been carried out are at each stage noted on the right, $C1 \leftrightarrow C2$ standing for the interchange of columns one and two, etc. The factor $(\lambda - 3)$ is chosen in the second operation since $(2\lambda^2 - 6\lambda + 1)/2\lambda = \lambda - 3 + 1/2\lambda$. For full details of the proof see Gantmacher (1959) or Marcus and Minc (1964).

Since all polynomials are over **C** it follows (as has been stated previously) that any polynomial can be written as a product of linear factors. Thus

$$i_j(\lambda) = (\lambda - \alpha_1)^{e_{j1}}(\lambda - \alpha_2)^{e_{j2}} \ldots (\lambda - \alpha_s)^{e_{js}}, j = 1, 2, \ldots, r$$

where each e_{jl} is a non-negative integer, and $(\lambda - \alpha_1), \ldots, (\lambda - \alpha_s)$ are the distinct linear factors of $i_r(\lambda)$ so that $e_{rl} > 0$, $l = 1, 2, \ldots, s$. Those factors

$$(\lambda - \alpha_k)^{e_{jl}}, \quad \text{all } k, j, l$$

which are not unity are called the *elementary divisors* of $A(\lambda)$.

Since i_j divides i_{j+1},

$$\cdots \leqslant e_{(j-1)1} \leqslant e_{j1} \leqslant e_{(j+1)1} \leqslant e_{(j+2)1} \cdots$$

$$\cdots \leqslant e_{(j-1)2} \leqslant e_{j2} \leqslant e_{(j+1)2} \leqslant \ldots, \text{etc.}$$

By this definition of elementary divisors it follows that Theorem 2–5–3 is equivalent to

Theorem 2–5–5

$A(\lambda)$, $B(\lambda)$ are equivalent if and only if they have the same rank and same elementary divisors.

For example, if a matrix has Smith form

$$\begin{bmatrix} \lambda^2 & 0 & 0 \\ 0 & \lambda^3(\lambda-1)^2 & 0 \\ 0 & 0 & \lambda^4(\lambda-1)^3(\lambda+1) \end{bmatrix} \tag{2–5–5}$$

then the determinantal divisors are $d_1 = \lambda^2$, $d_2 = \lambda^5(\lambda - 1)^2$, $d_3 = \lambda^9(\lambda - 1)^5 (\lambda + 1)$ and the elementary divisors are λ^2, λ^3, λ^4, $(\lambda - 1)^2$, $(\lambda - 1)^3$, $(\lambda + 1)$.

A direct relationship between similarity and equivalence is provided by

Theorem 2–5–6

Two square matrices A, B over **C** are similar if and only if their characteristic matrices $\lambda I - A$ and $\lambda I - B$ are equivalent over $\mathbf{C}[\lambda]$.

Theorem 2–5–7

If $A = \sum_{i=1}^{k} \oplus B_i$, where each B_i is a matrix of order n_i over **C**, then the set of elementary divisors of $\lambda I_n - A$ is the totality of elementary divisors of all the $\lambda I_{n_i} - B_i$, where $n = \sum_{i=1}^{k} n_i$.

If the Jordan form J of a matrix A is known then the elementary divisors of $\lambda I - A$ are easily written down, since by Theorem 2–5–6 they are the same as the elementary divisors of $\lambda I - J$. The elementary divisor of $\lambda I_k - J_k(\lambda_i)$ is just $(\lambda - \lambda_i)^k$ so by Theorem 2–5–7 the elementary divisors of $\lambda I - J$ are just the set of divisors of each Jordan block in J. For example, all matrices similar to

$$\begin{bmatrix} J_2(\lambda_1) & & \\ & J_1(\lambda_1) & \\ & & J_2(\lambda_2) \end{bmatrix} \tag{2–5–6}$$

will have $(\lambda - \lambda_1)^2$, $(\lambda - \lambda_1)$, $(\lambda - \lambda_2)^2$ as the elementary divisors of their characteristic matrix.

In fact Theorem 2–5–6 can be easily derived from a more general one concerning first degree polynomial matrices.

Theorem 2–5–8

If $A(\lambda)$, $B(\lambda)$ are two matrices equivalent over $\mathbf{C}[\lambda]$ with $A = \lambda A_1 - A_2$, $B = \lambda B_1 - B_2$, where A_1, A_2, B_1, B_2 are over $\mathbf{C}$ and if B is regular, then there exist *constant* non-singular matrices P, Q such that $PAQ = B$.

Because of their special importance the invariant factors of $\lambda I - A$, where A is a square matrix over $\mathbf{C}$ are often called the *similarity invariants* of A. It easily follows that the characteristic polynomial of A is the product of the similarity invariants, and that the elementary divisors of $\lambda I - A$ are powers of $\lambda - \lambda_i(A)$. Further:

Theorem 2–5–9

A is non-derogatory if and only if the first $n - 1$ similarity invariants are unity.

Theorem 2–5–9 is true in particular when the $\lambda_i(A)$ $(i = 1, 2, \ldots, n)$ are all distinct, in which case the elementary divisors of $\lambda I - A$ are all linear.

If the elementary divisors of a matrix $A(\lambda)$ of rank r are given, then $i_r(\lambda)$ is the product of the highest powers of factors $(\lambda - \alpha_k)$, $i_{r-1}(\lambda)$ is the product of highest powers in the remaining list of elementary divisors, and so on until all the elementary divisors have been used. The other invariant factors are then all unity. For example, the invariant factors of the matrix in (2–5–5) are obtained as $i_3 = \lambda^4(\lambda - 1)^3(\lambda + 1)$, $i_2 = \lambda^3(\lambda - 1)^2$, $i_1 = \lambda^2$; the invariant factors of the matrix in (2–5–6) are $i_5 = (\lambda - \lambda_1)^2(\lambda - \lambda_2)^2$, $i_4 = (\lambda - \lambda_1)$, $i_3 = i_2 = i_1 = 1$.

The roots of

$$\det A(\lambda) = 0 \tag{2–5–7}$$

where A is an nth-order matrix of degree N are often required in applications but calculation is generally difficult; if A is regular (2–5–7) has Nn roots. A description of some numerical methods, with particular reference to vibration systems, is given by Lancaster (1966).

It is worth pointing out that since $\mathbf{C}[\lambda]$ and the set of integers are both examples of an integral domain [Birkhoff and MacLane (1965)] many of the results of this section carry over immediately to the case of matrices with integer elements.

2–6 Rational matrices

A *rational* $m \times n$ matrix $Z(\lambda) = (z_{ij}(\lambda))$ is one whose elements are rational functions of λ with coefficients in $\mathbf{C}$, so

$$z_{ij}(\lambda) = x_{ij}(\lambda)/y_{ij}(\lambda)$$

where $x_{ij}(\lambda)$, $y_{ij}(\lambda)$ are polynomials. Let $z(\lambda)$ be the (monic) least common multiple of all the y_{ij}. Then zZ is a polynomial matrix with Smith form (Section 2–5)

$$\begin{bmatrix} i_1 & & & & & & & \\ & \cdot & & & & & & \\ & & \cdot & & & & & \\ & & & \cdot & & & & \\ & & & & i_R & & & \\ & & & & & 0 & & \\ & & & & & & \ddots & \\ & & & & & & & 0 \end{bmatrix}$$

where the i_j are the invariant factors of zZ and $R = \text{rank}\,(zZ)$. It then follows that:

Theorem 2–6–1
Every rational matrix $Z(\lambda)$ can be expressed as

$$Z = PMQ$$

where $P(\lambda)$, $Q(\lambda)$ are invertible polynomial matrices and

$$M = \begin{bmatrix} \phi_1/\psi_1 & & & & & & & & & \\ & \cdot & & & & & & & & \\ & & \cdot & & & & & & & \\ & & & \cdot & & & & & & \\ & & & & \phi_r/\psi_r & & & & & \\ & & & & & \phi_{r+1} & & & & \\ & & & & & & \ddots & & & \\ & & & & & & & \phi_R & & \\ & & & & & & & & 0 & \\ & & & & & & & & & \ddots \\ & & & & & & & & & & 0 \end{bmatrix}$$

is the *Smith-McMillan form* [McMillan (1952)] of $Z(\lambda)$ satisfying the conditions:

(i) $\phi_j(\lambda)$, $\psi_j(\lambda)$ are relatively prime monic polynomials;
(ii) $\phi_j/\psi_j = i_j(zZ)/z$;
(iii) each $\phi_j(\lambda)$ is a factor of $\phi_{j+1}(\lambda)$, and $\phi_1(\lambda) = i_1(zZ)$;
(iv) each $\psi_j(\lambda)$ is a factor of $\psi_{j-1}(\lambda)$ and $\psi_1(\lambda) = z(\lambda)$.

Rational matrices play an important part in network theory [Newcomb (1966)], realization theory [Kalman (1965)], and linear system theory [Kalman (1966), Rosenbrock (1967)].

2–7 Quadratic and hermitian forms

A *real quadratic form* is the polynomial

$$\sum_{i=1}^{n}\sum_{j=1}^{n} a_{ij}x_i x_j \tag{2–7–1}$$

in the n real variables $x_1, x_2, \ldots, x_n$. The expression (2–7–1) can be written

$$x'Ax \tag{2–7–2}$$

where $x = (x_1, x_2, \ldots, x_n)'$ and the real matrix $A = (a_{ij})$ is called the *matrix of the form*. There is no loss of generality in taking A to be symmetric. For suppose A is not symmetric, and write

$$\begin{aligned} A &= \tfrac{1}{2}(A + A') + \tfrac{1}{2}(A - A') \\ &= A_1 + A_2, \end{aligned}$$

so that A_1 is symmetric and A_2 skew-symmetric.

Then
$$\begin{aligned} x'A_2x &= (x'A_2x)' \\ &= -x'A_2x \quad \text{(Theorem 1–3–1)}, \end{aligned}$$

whence $x'A_2x = 0$ and $x'Ax = x'A_1x$. The *rank* of (2–7–1) is defined to be $R(A)$.

If the *linear transformation*

$$y = Px, \tag{2–7–3}$$

where y is also an n-vector and P is non-singular, is applied to (2–7–2) the quadratic form becomes $y'P'APy$, so that the matrix of the form is $B = P'AP$; i.e., B and A are congruent matrices. By Theorem 1–13–6 it follows that:

Theorem 2–7–1
There exists a matrix P such that B is diagonal, so that (2–7–2) is reduced to a sum of squares $\sum_{i=1}^{r} \alpha_i y_i^2$, where $r = R(A)$ and each $\alpha_i \neq 0$. Furthermore, there exists a P which makes each α_i equal to ± 1 and an orthogonal P which makes

$$\alpha_i = \lambda_i(A),\ i = 1, 2, \ldots, r.$$

More generally,

Theorem 2–7–2 (Sylvester's law of inertia)
If the form (2–7–2) is reduced by two real non-singular linear transformations to $\sum_{i=1}^{r} \alpha_i y_i^2$ and $\sum_{i=1}^{r} \beta_i z_i^2$ then the number of positive α_i is equal to the number of positive β_i.

The number of positive α_i is called the *index* of the form or of its matrix, and the *signature* is equal to the index minus the number of negative α_i. Notice that Theorem 2–7–1 implies that the index is equal to the number of positive $\lambda_i(A)$.

Theorem 2–7–3
Two real symmetric matrices are congruent with real transforming matrix if and only if they have the same rank and index.

The generalization of real quadratic forms is not to complex forms but to hermitian forms

$$\sum_{i=1}^{n} \sum_{j=1}^{n} h_{ij}\bar{x}_i x_j = x^* H x \tag{2–7–4}$$

where H is a hermitian matrix. For any values of x_i the form (2–7–4) takes only real values, since $\overline{(x^* Hx)} = x' \bar{H} \bar{x} = (x^* H^* x)' = x^* Hx$. Indeed if the x_i take only real values then (2–7–4) reduces to (2–7–2), so a quadratic form is just a special case of a hermitian form (with real matrix). Results for hermitian forms corresponding to those for quadratic forms may now be stated.

Two square matrices L, M over **C** are called *hermitely congruent* or *conjunctive* if there exists a non-singular P over **C** such that $L = P^*MP$. This definition is needed since the transformation corresponding to (2–7–3) is now complex and makes (2–7–4) into y^*P^*HPy.

Theorem 2–7–1a
If H is of rank r then there exists a unitary P which reduces (2–7–4) to $\sum_{i=1}^{r} \lambda_i \bar{x}_i x_i$, where $\lambda_1, \ldots, \lambda_r$ are the (real) non-zero characteristic roots of H.

As before,

Theorem 2–7–2a
If (2–7–4) is reduced to $\sum \alpha_i \bar{y}_i y_i$ and $\sum \beta_i \bar{z}_i z_i$ then the number of positive α (the *index* of the form) is the same as the number of positive β_i.

Theorem 2–7–3a
Two hermitian matrices are conjunctive if and only if they have the same rank and index.

An important concept which can be introduced here is that of the *inertia* of an $n \times n$ complex matrix A, defined to be the integer triple $(\pi(A), \nu(A), \delta(A))$, where $\pi(A)$, $\nu(A)$, and $\delta(A)$ are the numbers of characteristic roots of A with positive, negative, and zero real parts respectively. Theorem 2–7–2a can then be written as

Theorem 2–7–2b (Sylvester conditions)
If H is hermitian and P non-singular,

$$In\, H = In\,(P^*HP).$$

Another interesting theorem on the inertia of hermitian matrices which is easily proved depends upon the fact (Theorem 2–1–3) that for any hermitian matrix H of rank at least k there exists a (real) permutation matrix R such that

$$J = R'HR = \begin{bmatrix} H_{11} & H_{12} \\ H_{12}^* & H_{22} \end{bmatrix} \tag{2–7–5}$$

with H_{11} of order k and non-singular. By virtue of Theorem 2–7–2*b*, it can be assumed that any hermitian matrix J is in the form (2–7–5). Then

Theorem 2–7–4 [*Haynsworth* (*1968*)]

$$In\, J = In\, H_{11} + In\, K, \text{ where } K = H_{22} - H_{12}^* H_{11}^{-1} H_{12}.$$

Proof. Let

$$L = \begin{bmatrix} I_k & -H_{11}^{-1} H_{12} \\ 0 & I_{n-k} \end{bmatrix}.$$

It is easily verified that

$$L^*JL = \begin{bmatrix} H_{11} & 0 \\ 0 & K \end{bmatrix}.$$

Since $In\, L^*JL = In\, J$, the result follows (using Theorem 2–3–13).

Notice also that by Schur's formula, Theorem 1–10–1,

$$\det J = \det H_{11} \det K.$$

Other results on inertia may be found in Ostrowski and Schneider (1962), Carlson and Schneider (1963), Carlson (1965), and Haynsworth and Ostrowski (1968).

An interesting application of quadratic forms is to the estimation of the characteristic roots of a symmetric matrix.

Theorem 2–7–5 (*Rayleigh's principle*)
If λ_1, λ_2 are the greatest and least roots of A then

$$\lambda_1 \geqslant \frac{x'Ax}{x'x} \geqslant \lambda_2,$$

for any (real) vector x. That is,

$$\lambda_1 = \max_x\, [x'Ax/x'x], \quad \lambda_2 = \min_x\, [x'Ax/x'x].$$

The result of Theorem 2–7–5 has been generalized by Courant and Fischer to provide variational formulae for all the $\lambda_i(A)$ [Bellman (1960), p. 113].

Exercise 2–7–1
If A is any real $m \times n$ matrix show that $x'A'Ax \geqslant 0$ for any real column n-vector x. What is the corresponding result if A is complex?

Exercise 2–7–2

If A is symmetric, S skew-symmetric, and $(A + S)$ non-singular show that the matrix $P = (A + S)^{-1}(A - S)$ satisfies the relations $P'(A + S)P = A + S, P'(A - S)P = A - S$ and deduce that $x'Ax$ is unaltered by the transformation $x = Py$.

2–8 Sign definite forms

Important classes of quadratic or hermitian forms $\phi(x)$ are those which do not change sign:

(i) ϕ is said to be *positive* (*negative*) *definite* if $\phi > 0$ (<0) except when $x = 0$;
(ii) ϕ is *positive* (*negative*) *semi-definite* if $\phi \geqslant 0$ $(\leqslant 0)$ for all values of x;
(iii) otherwise ϕ is *indefinite*.

The descriptions of (i), (ii), and (iii) are also applied to the corresponding matrix of the form ϕ. Notice that a definite form is also semi-definite.

A basic problem is to determine to which category a given form belongs. The tests given below apply equally to hermitian or quadratic forms $\phi = x^*Hx$ and thus to hermitian or real symmetric matrices.

Theorem 2–8–1

ϕ is positive (negative) definite if and only if all $\lambda_i(H) > 0$ (< 0).

Theorem 2–8–2

ϕ is positive (negative) semi-definite if and only if all $\lambda_i(H) \geqslant 0$ $(\leqslant 0)$.

Theorem 2–8–3

ϕ is indefinite if and only if at least one $\lambda_i(H)$ is negative and one positive.

These results can be expressed in a way which does not require calculation of $\lambda_i(H)$ but merely reduction of ϕ to diagonal form, which can be achieved fairly easily by Lagrange's method [Mirsky (1963), p. 371].

Theorem 2–8–4

Let r, s be the rank and signature of ϕ.

Then: ϕ is positive (negative) definite if and only if $r = s = n$ $(r = -s = n)$;
ϕ is positive (negative) semi-definite if and only if $r = s \leqslant n$ $(r = -s \leqslant n)$;
ϕ is indefinite if and only if $|s| < r$.

Determinantal criteria for definiteness are of simple form:

Theorem 2–8–5

Denote by D_i the minors det $H(1, 2, \ldots, i \mid 1, 2, \ldots, i)$.

Then: ϕ is positive definite if and only if $D_i > 0$, all i
negative definite if and only if $D_i > 0$, all even i
$D_i < 0$, all odd i
positive semi-definite if and only if $|H| = 0$, $D_i \geqslant 0$, all i
negative semi-definite if and only if $|H| = 0$, $D_i \geqslant 0$, all even i
$D_i \leqslant 0$, all odd i
indefinite if at least one of the following conditions holds:
(*a*) $D_i < 0$ for at least one even i
(*b*) $D_i > 0$, $D_j < 0$ for at least one pair i, j both odd.

The disadvantage in applying the tests of Theorem 2–8–5 in practice is, of course, due to the difficulty of evaluating large order determinants. A more convenient numerical procedure is provided by:

Theorem 2–8–6
H is positive definite hermitian if and only if it can be expressed as P^*P for some non-singular matrix P; in other words if and only if it is hermitely congruent to I_n. In particular P may be taken as a triangular matrix in which case computation of the elements of P is very simple.

For example, when H is real symmetric it is positive definite if and only if P is real non-singular.

Let $H = (h_{ij})$, $P = (p_{ij})$ then $H = P'P$ gives

$$p_{11} = \sqrt{h_{11}},\ p_{ii} = \sqrt{(h_{ii} - p_{1i}^2 - p_{2i}^2 \cdots - p_{i-1,i}^2)},\ i = 2, \ldots, n$$

$$p_{jk} = \frac{h_{jk} - \sum_{i=1}^{j-1} p_{ij}\, p_{ik}}{p_{jj}} \qquad \begin{array}{l} j = 1, 2, \ldots, n \\ k = j+1, \ldots, n. \end{array}$$

$$p_{jk} = 0,\ j > k.$$

Thus P is upper triangular and can be found easily row by row. The condition for non-singularity is just $p_{ii} > 0$.

An important generalization of Theorem 2–8–6 is the fact that sign definiteness of a hermitian form is unaltered by non-singular linear transformation:

Theorem 2–8–7
If H is positive (negative) definite hermitian then so is P^*HP for any non-singular matrix P.

Finally, a result concerning a pair of hermitian forms x^*Hx, x^*Jx:

Theorem 2–8–8
If J is positive definite then there exists a non-singular linear transformation (2–7–3) over **C** which simultaneously reduces x^*Hx to $\sum_{i=1}^{n} \lambda_i y_i \bar{y}_i$ and x^*Jx to $\sum_{i=1}^{n} y_i \bar{y}_i$ where the λ_i are the roots of

$$|H - \lambda J| = 0.$$

These roots are all real but not necessarily all distinct. In the case of quadratic forms the matrix of the linear transformation is real. If H, J are both positive definite λ_i are all positive.

Exercise 2–8–1

Let $\phi(x) = x'Ax$ be a positive semi-definite quadratic form and let $\psi(x,y) = x'Ay$ be the associated *bilinear* form (y is a column n-vector). Show that, for all real t, x, y

$$\phi(x + ty) = t^2\phi(y) + 2t\psi(x,y) + \phi(x) \geqslant 0.$$

Hence deduce the *Cauchy-Schwarz* inequality

$$[\psi(x,y)]^2 \leqslant \phi(x)\,\phi(y).$$

[Hint: regard as a quadratic in t and complete the square.] Write down the result when $A = I_n$.

Exercise 2–8–2

A simple proof that the roots of a hermitian matrix H (and hence those of a real symmetric matrix) are all real is as follows: Let x, λ be corresponding characteristic vector and root, so that $Hx = \lambda x$. Therefore $x^*Hx = \lambda x^*x$. Take the conjugate transpose of this last expression to get $x^*H^*x = \bar{\lambda}x^*x$. Since $H^* = H$, subtraction gives $\bar{\lambda}x^*x = \lambda x^*x$, whence $\bar{\lambda} = \lambda$ since $x^*x \neq 0$ if $x \neq 0$.

A real matrix A is called *symmetrizable* if there exists a positive definite matrix M such that $A'M = MA$ [Bellman (1960)]. Using a similar method to that given above show that all the characteristic roots of A are real.

Exercise 2–8–3

Use the method of Exercise 2–8–2 to show that if H, J are hermitian and K is skew-hermitian then the roots of $H^{-1}J$ are all real and those of $H^{-1}K$ are either zero or purely imaginary.

2–9 Norms

A norm is a single scalar which provides a measure of the 'size' of a matrix or vector. The *euclidean norm* of a vector over **C** is just what is normally understood as the length of the vector, namely

$$\begin{aligned} \| x \|_e &= [\sum_{i=1}^{n} | x_i |^2]^{1/2} \\ &= [xx^*]^{1/2}, \end{aligned} \tag{2–9–1}$$

where $x = (x_1, \ldots, x_n)$ and $| x_i |$ denotes modulus of x_i. The generalization to the euclidean norm of a matrix has already been given (Theorem 1–8–3) as

$$\begin{aligned} \| A \|_e &= [\sum_{i=1}^{m} \sum_{j=1}^{n} | a_{ij} |^2]^{1/2} \\ &= [tr\, AA^*]^{1/2}. \end{aligned} \tag{2–9–2}$$

Other norms are also used, for example

$$\sum_{i=1}^{n} | x_i | \tag{2–9–3}$$

as an alternative vector norm, and

$$(\text{maximum characteristic root of } A^*A)^{1/2}, \tag{2–9–4}$$

or

$$\max_{i,j} | a_{ij} | \quad \text{(the cubic norm)} \tag{2–9–5}$$

as alternative matrix norms. For more examples and discussion see Wilkinson (1965).

All norms for both matrices and vectors over **C** satisfy the conditions

$$\| X \| > 0 \quad \text{unless } X = 0$$

$$\| kX \| = | k | \, \| X \| \quad \text{for any scalar } k \text{ in } \mathbf{C}$$

$$\| X + Y \| \leqslant \| X \| + \| Y \| \tag{2–9–6}$$

$$\| XY \| \leqslant \| X \| \| Y \|. \text{ (This, of course, only holds for matrices.)}$$

If any matrix and vector norms are such that

$$\| Ax \| \leqslant \| A \| \, \| x \| \tag{2–9–7}$$

for any A, x then the two norms are said to be *consistent* (or *compatible*). In particular the euclidean vector norm (2–9–1) and matrix norm (2–9–2) are consistent, and the norms (2–9–1) and (2–9–4) are also consistent.

If $\| Ax \|$ is any vector norm and a matrix norm $\| A \|_s$ is defined by

$$\| A \|_s = \max_{\|x\|=1} \| Ax \|$$

then the matrix norm $\| A \|_s$ is said to be *subordinate* to the given vector norm.

Theorem 2–9–1

If $\| A \|$ is a matrix norm consistent with a vector norm $\| x \|$, then $\| A \|_s \leqslant \| A \|$.

Theorem 2–9–2

If λ is a characteristic root of A and x a corresponding vector then for consistent norms, $Ax = \lambda x$ implies, using (2–9–6) and (2–9–7) that

$$| \lambda | \leqslant \| A \|. \tag{2–9–8}$$

Defining the *spectral radius* of A as

$$\rho(A) = \max_i | \lambda_i(A) |,$$

Eqn (2–9–8) gives $\rho(A) \leqslant \| A \|$. That is,

Theorem 2–9–3

The spectral radius of a given matrix is less than or equal to the value of any of the norms.

There is an important connection with convergence of a matrix sequence (see Chapter 4).

Theorem 2–9–4
A sufficient condition for $A^k \to 0$ as $k \to \infty$ is that $\| A \| < 1$ for any norm. A necessary and sufficient condition is that $\rho(A) < 1$.

Theorem 2–9–5
More generally, $(A')^k C A^k \to 0$ as $k \to \infty$, for any constant matrix C if and only if $\rho(A) < 1$.

A very full account of norms and related topics, with many references, is given in Chapter 2 of Householder (1964). [See also Faddeev and Faddeeva (1963).]

Exercise 2–9–1
Show that the matrix norm subordinate to the euclidean vector norm is that given by (2–9–4) (use Theorem 2–7–5).

2–10 Some matrix equations

Theorem 2–10–1
If $f(\lambda) = (\lambda - \lambda_1)^{\alpha_1}(\lambda - \lambda_2)^{\alpha_2} \cdots (\lambda - \lambda_h)^{\alpha_h}$ then the solution of the matrix equation $f(X) = 0$, where X is square of order n, is

$$X = T \operatorname{diag}\,[\lambda_{i_1} I_{\beta_{i_1}} + J_{\beta_{i_1}}(0), \ldots, \lambda_{i_k} I_{\beta_{i_k}} + J_{\beta_{i_k}}(0)]T^{-1}, \qquad (2\text{–}10\text{–}1)$$

where T is an arbitrary non-singular matrix of order n; $i_1, i_2, \ldots, i_k$ are integers from the set $1, 2, \ldots, h$ (some of the i_j may be equal); $\beta_{i_1} \leqslant \alpha_{i_1}, \beta_{i_2} \leqslant \alpha_{i_2}, \ldots, \beta_{i_k} \leqslant \alpha_{i_k}$ and $\beta_{i_1} + \beta_{i_2} + \cdots + \beta_{i_k} = n$.

For example, an idempotent matrix satisfies the equation $X^2 = X$, so $f(\lambda) = \lambda(\lambda - 1)$. Thus $\alpha_1 = \alpha_2 = 1$ and the only possible contributions to the diagonal matrix in (2–10–1) are $J_\beta(0)$, $I_\gamma + J_\gamma(0)$, with $\beta \leqslant 1, \gamma \leqslant 1$. Thus the general expression for an idempotent matrix is

$$X = T \operatorname{diag}\,(1, 1, \ldots, 1, 0, 0, \ldots, 0)T^{-1},$$

where the number of unit elements in the diagonal matrix is arbitrary.

Similarly, the general solution of $X^2 = 0$ is

$$X = T \operatorname{diag}\,\left(0, 0, \ldots, 0, \begin{bmatrix} 0 & 1 \\ 0 & 0 \end{bmatrix}, \ldots\right)T^{-1},$$

the sequence and number of the diagonal entries of each type again being entirely arbitrary.

Theorem 2–10–2
The equation

$$AX - XB = C, \qquad (2\text{–}10\text{–}2)$$

where A is $m \times m$, B is $n \times n$, and X is $m \times n$, has a unique solution if and only if A and B have no common characteristic roots.

It is of value to give a proof of this theorem, since the method employed will be useful later.

Using Theorem 1–6–4, Eqn (2–10–2) can be written

$$(A \otimes I_n - I_m \otimes B')x = c, \tag{2–10–3}$$

x and c being column mn-vectors formed from the rows of X, C respectively, taken in order. By Theorem 1–12–1, (2–10–3) has a unique solution if and only if the matrix on the left in (2–10–3) is non-singular; but by Theorem 2–3–15 this matrix has characteristic roots $\lambda_i(A) - \lambda_j(B')$ and so is non-singular if and only if no two $\lambda_i(A)$, $\lambda_j(B)$ are equal.

An important particular case of Theorem 2–10–2 is

Theorem 2–10–3

The equation

$$AX = XB \tag{2–10–4}$$

has a non-zero solution if and only if A and B have at least one common characteristic root.

Also important is the even more special case when $A = B$.

Theorem 2–10–4 (*commuting matrices*)

The number of linearly independent solutions for the nth-order matrix X of the equation

$$AX = XA \tag{2–10–5}$$

is $\delta_n + 3\delta_{n-1} + 5\delta_{n-2} + \cdots + (2n-1)\delta_1$ where $\delta_1, \delta_2, \ldots, \delta_n$ are the degrees of the invariant polynomials of $\lambda I_n - A$. Further, every solution X is a polynomial in A if and only if A is non-derogatory.

Full details of the general algebraic expressions for the solutions of (2–10–2, 4, 5) are given by Gantmacher (1959).

Exercise 2–10–1

Show that the general expression for a nilpotent matrix of index r (see Exercise 2–2–1) is

$X = T\,\mathrm{diag}\,[J_{\beta_1}(0), J_{\beta_2}(0), \ldots, J_{\beta_k}(0)]T^{-1}$ where $\beta_1, \beta_2, \ldots, \beta_k \leqslant r$ and $\sum_{j=1}^{k} \beta_j = n$.

Exercise 2–10–2

Show that the number of linearly independent solutions of (2–10–5) can also be written as $n + 2\sum_{i=1}^{n-1} m_i$, where m_i is the degree of the ith-determinantal divisor of $\lambda I_n - A$.

Discussion and further reading

This chapter has covered more advanced areas of matrix theory, but again the aim has been primarily to present a useful compendium rather than detailed proofs. Most of the classical results on characteristic roots, polynomial matrices, matrix equations, etc., can be found in full in such texts as Gantmacher (1959) or Cullen (1966) but a few, for example the Smith-McMillan form, are not so easily accessible. An extremely comprehensive and critical description of methods for calculation of characteristic roots and vectors is contained in Wilkinson (1965); incidentally this book begins with a brief survey of many of the topics covered in this chapter. Lancaster (1966) deals with more specialized problems of polynomial matrices.

References

BELLMAN, R. (1960), *Introduction to matrix analysis*, McGraw-Hill, New York.

BIRKHOFF, G. and S. MACLANE (1965), *A survey of modern algebra*, 3rd edn., Macmillan, New York.

CARLSON, D. (1965), Rank and inertia bounds for matrices under $R(AH) \geqslant 0$, *J. Math. Anal. Applics*, **10,** 100–111.

CARLSON, D. and H. SCHNEIDER (1963), Inertia theorems for matrices: the semi-definite case, *J. Math. Anal. Applics*, **6,** 430–446.

CULLEN, G. C. (1966), *Matrices and linear transformations*, Addison-Wesley, Reading, Massachusetts.

FADDEEV, D. K. and V. N. FADDEEVA (1963), *Computational methods of linear algebra*, W. H. Freeman, San Francisco.

GANTMACHER, F. R. (1959), *The theory of matrices*, Vol. I, Chelsea Publishing Co. New York.

HAYNSWORTH, E. V. (1968), Determination of the inertia of a partitioned Hermitian matrix, *Linear Algebra*, **1,** 73–81.

HAYNSWORTH, E. V. and A. M. OSTROWSKI (1968), On the inertia of some classes of partitioned matrices, *Linear Algebra*, **1,** 299–316.

HODGE, W. V. D. and D. PEDOE (1947), *Methods of algebraic geometry*, Vol. I, Cambridge University Press, London.

HOUSEHOLDER, A. S. (1964), *The theory of matrices in numerical analysis*, Blaisdell, Waltham, Massachusetts.

KALMAN, R. E. (1965), Irreducible realizations and the degree of a rational matrix, *SIAM J. Appl. Math.*, **13,** 520–544.

KALMAN, R. E. (1966), On structural properties of linear, constant, multivariable systems, I.F.A.C. Conference, London, paper 6A.

LANCASTER, P. (1966), *Lambda-matrices and vibrating systems*, Pergamon Press, Oxford.

MARCUS, M. and H. MINC (1964), *A survey of matrix theory and matrix inequalities*, Allyn and Bacon, Boston.

McMILLAN, B. (1952), Introduction to formal realizability theory, *Bell System Tech. J.*, **31,** 541–600.

MIRSKY, L. (1963), *An introduction to linear algebra*, Oxford University Press, London.

NEWCOMB, R. W. (1966), *Linear multiport synthesis*, McGraw-Hill, New York.

OSTROWSKI, A. and H. SCHNEIDER (1962), Some theorems on the inertia of general matrices, *J. Math. Anal. Applics*, **4,** 72–84.

ROSENBROCK, H. H. (1967), On linear system theory, *Proc. I.E.E.*, **114**, 1353–1359.
SCHNEIDER, H. (1964), *Recent advances in matrix theory*, University of Wisconsin Press.
WILKINSON, J. H. (1965), *The algebraic eigenvalue problem*, Oxford University Press, London.

3

Special forms of matrix

3–1 The Jordan normal form

The Jordan form was introduced in Section 2–4 and some of its basic properties are described there. In fact, if the elementary divisors of an $n \times n$ matrix A (over the complex field) are

$$(\lambda - \lambda_1)^{e_1}, (\lambda - \lambda_2)^{e_2}, \ldots, (\lambda - \lambda_r)^{e_r}$$

its Jordan form is the matrix

$$J = \text{diag}\,(J_1, J_2, \ldots, J_r)$$

where the $e_i \times e_i$ matrix J_i is

$$J_i = \begin{bmatrix} \lambda_i & 1 & 0 & \cdots & 0 \\ 0 & \lambda_i & 1 & \cdots & 0 \\ \cdot & \cdot & \cdot & \cdots & \cdot \\ \cdot & \cdot & \cdot & \cdots & \cdot \\ \cdot & \cdot & \cdot & \cdots & \cdot \\ 0 & 0 & 0 & \cdots & 1 \\ 0 & 0 & 0 & \cdots & \lambda_i \end{bmatrix}$$

for $i = 1, 2, \ldots, r$. A and J are similar by Theorem 2–4–7 and, of course, not all of the λ_i need be distinct.

Example 3–1–1

Consider the powers of a 3×3 Jordan block

$$J_3^2 = \begin{bmatrix} \lambda & 1 & 0 \\ 0 & \lambda & 1 \\ 0 & 0 & \lambda \end{bmatrix} \begin{bmatrix} \lambda & 1 & 0 \\ 0 & \lambda & 1 \\ 0 & 0 & \lambda \end{bmatrix} = \begin{bmatrix} \lambda^2 & 2\lambda & 1 \\ 0 & \lambda^2 & 2\lambda \\ 0 & 0 & \lambda^2 \end{bmatrix}$$

$$J_3^3 = \begin{bmatrix} \lambda & 1 & 0 \\ 0 & \lambda & 1 \\ 0 & 0 & \lambda \end{bmatrix} \begin{bmatrix} \lambda^2 & 2\lambda & 1 \\ 0 & \lambda^2 & 2\lambda \\ 0 & 0 & \lambda^2 \end{bmatrix} = \begin{bmatrix} \lambda^3 & 3\lambda^2 & 3\lambda \\ 0 & \lambda^3 & 3\lambda^2 \\ 0 & 0 & \lambda^3 \end{bmatrix}$$

$$J_3^4 = \begin{bmatrix} \lambda & 1 & 0 \\ 0 & \lambda & 1 \\ 0 & 0 & \lambda \end{bmatrix} \begin{bmatrix} \lambda^3 & 3\lambda^2 & 3\lambda \\ 0 & \lambda^3 & 3\lambda^2 \\ 0 & 0 & \lambda^3 \end{bmatrix} = \begin{bmatrix} \lambda^4 & 4\lambda^3 & 6\lambda^2 \\ 0 & \lambda^4 & 4\lambda^3 \\ 0 & 0 & \lambda^4 \end{bmatrix}$$

and so on. In general if we write

$$J_3^{r-1} = \begin{bmatrix} \lambda^{r-1} & \binom{r-1}{1}\lambda^{r-2} & \binom{r-1}{2}\lambda^{r-3} \\ 0 & \lambda^{r-1} & \binom{r-1}{1}\lambda^{r-2} \\ 0 & 0 & \lambda^{r-1} \end{bmatrix}$$

then

$$J_3^{r} = \begin{bmatrix} \lambda & 1 & 0 \\ 0 & \lambda & 1 \\ 0 & 0 & \lambda \end{bmatrix} \begin{bmatrix} \lambda^{r-1} & \binom{r-1}{1}\lambda^{r-2} & \binom{r-1}{2}\lambda^{r-3} \\ 0 & \lambda^{r-1} & \binom{r-1}{1}\lambda^{r-2} \\ 0 & 0 & \lambda^{r-1} \end{bmatrix}$$

$$= \begin{bmatrix} \lambda^{r} & \binom{r}{1}\lambda^{r-1} & \binom{r}{2}\lambda^{r-2} \\ 0 & \lambda^{r} & \binom{r}{1}\lambda^{r-1} \\ 0 & 0 & \lambda^{r} \end{bmatrix}$$

since $$\binom{r-1}{1} + 1 = \frac{(r-1)!}{(r-2)!} + 1 = \binom{r}{1}$$

$$\binom{r-1}{1} + \binom{r-1}{2} = \frac{(r-1)!}{(r-2)!} + \frac{(r-1)!}{(r-3)!2!} = \frac{r!}{(r-2)!2!} = \binom{r}{2}.$$

Thus by induction the above result is true for any positive integer r.

Exercise 3–1–1

Use induction in the same way as above to show that the nth power of a general $k \times k$ Jordan block is given by

$$J^n = \begin{bmatrix} \lambda^n & \binom{n}{1}\lambda^{n-1} & \cdots & \binom{n}{k-1}\lambda^{n-k+1} \\ 0 & \lambda^n & \cdots & \binom{n}{k-2}\lambda^{n-k+2} \\ \cdot & \cdot & \cdots & \cdots \\ \cdot & \cdot & \cdots & \cdots \\ \cdot & \cdot & \cdots & \cdots \\ 0 & 0 & \cdots & \lambda^n \end{bmatrix}.$$

We now discuss briefly the similarity transformation from an arbitrary $n \times n$ matrix A into its Jordan form J. If T is the required transformation matrix then

$$J = T^{-1}AT \tag{3–1–1}$$

or

$$TJ = AT \tag{3–1–2}$$

giving a set of n^2 linear equations for the unknown elements of T. Although for a given A the matrix J is uniquely determined except for the order of its diagonal blocks there are infinitely many transforming matrices T. This follows from the fact that Eqn (3–1–2) is equally satisfied by $Q = UT$, where U is non-singular and such that $UA = AU$ but otherwise arbitrary.

It is obvious that direct solution of the linear system (3–1–2) of n^2 equations is not efficient unless T has some special form (triangular, tridiagonal, etc.) which reduces the computation to a manageable proportion. The following method [Gantmacher (1959)] avoids direct solution of (3–1–2).

Bring the characteristic matrix $\lambda I - A$ into its Smith form by a sequence of elementary operations of which those concerning columns are $E_1, E_2, \ldots, E_r$. Determine the elementary divisors and hence the Jordan form J of A (Sections 2–4 and 2–5). Reduce the characteristic matrix $\lambda I - J$ to its Smith form by a series of elementary transformations of which those concerning columns are $G_1, G_2, \ldots, G_s$. Now form the polynomial matrix $P(\lambda)$ by operating on the unit matrix I_n with the elementary column transformations

$$E_1E_2 \cdots E_rG_s^{-1}G_{s-1}^{-1} \cdots G_1^{-1}.$$

Finally, the transforming matrix T is obtained by writing $P(\lambda)$ as a matrix polynomial and substituting J for λ.

Example 3–1–2

$$\text{Let } A = \begin{bmatrix} 0 & 1 & 1 \\ 0 & 2 & 0 \\ -2 & 1 & 3 \end{bmatrix}.$$

We first find the Smith form of $\lambda I - A$ by using elementary transformations. Each transformation is written by the side of the matrix and the notation of Section 2–5 is adopted. Notice that in actually carrying out a reduction to Smith form the full procedure outlined in Section 2–5 need not always be used, short cuts often being seen by inspection.

$$\lambda I - A = \begin{bmatrix} \lambda & -1 & -1 \\ 0 & \lambda - 2 & 0 \\ 2 & -1 & \lambda - 3 \end{bmatrix}$$

$$\sim \begin{bmatrix} -1 & -1 & \lambda \\ 0 & \lambda - 2 & 0 \\ \lambda - 3 & -1 & 2 \end{bmatrix} \quad C_1 \leftrightarrow C_3$$

$$\sim \begin{bmatrix} -1 & -1 & 0 \\ 0 & \lambda - 2 & \lambda^2 - 2\lambda \\ \lambda - 3 & -1 & 2 - \lambda \end{bmatrix} \quad C_3 + \lambda C_2$$

$$\sim \begin{bmatrix} -1 & 0 & -1 \\ 0 & \lambda^2 - 2\lambda & \lambda - 2 \\ \lambda - 3 & 2 - \lambda & -1 \end{bmatrix} \quad C_2 \leftrightarrow C_3$$

$$\sim \begin{bmatrix} -1 & 0 & 0 \\ 0 & \lambda^2 - 2\lambda & \lambda - 2 \\ \lambda - 3 & 2 - \lambda & 2 - \lambda \end{bmatrix} \quad C_3 - C_1$$

$$\sim \begin{bmatrix} -1 & 0 & 0 \\ 0 & \lambda^2 - 3\lambda + 2 & \lambda - 2 \\ \lambda - 3 & 0 & 2 - \lambda \end{bmatrix} \quad C_2 - C_3$$

$$\sim \begin{bmatrix} -1 & 0 & 0 \\ 0 & \lambda^2 - 3\lambda + 2 & \lambda - 2 \\ 0 & 0 & 2 - \lambda \end{bmatrix} \quad R_3 + (\lambda - 3)R_1$$

$$\sim \begin{bmatrix} 1 & 0 & 0 \\ 0 & (\lambda - 2)(\lambda - 1) & \lambda - 2 \\ 0 & 0 & 2 - \lambda \end{bmatrix} \quad (-1)C_1$$

$$\sim \begin{bmatrix} 1 & 0 & 0 \\ 0 & \lambda - 2 & (\lambda - 2)(\lambda - 1) \\ 0 & 2 - \lambda & 0 \end{bmatrix} \quad C_2 \leftrightarrow C_3$$

$$\sim \begin{bmatrix} 1 & 0 & 0 \\ 0 & \lambda - 2 & 0 \\ 0 & 2 - \lambda & (\lambda - 1)(\lambda - 2) \end{bmatrix} \quad C_3 - (\lambda - 1)C_2$$

$$\sim \begin{bmatrix} 1 & 0 & 0 \\ 0 & \lambda - 2 & 0 \\ 0 & 0 & (\lambda - 1)(\lambda - 2) \end{bmatrix} \quad R_3 + R_2.$$

It follows that the elementary divisors of A are $(\lambda - 1)$, $(\lambda - 2)$, $(\lambda - 2)$, and the Jordan form of A is diagonal in this case and given by

$$J = \begin{bmatrix} 1 & 0 & 0 \\ 0 & 2 & 0 \\ 0 & 0 & 2 \end{bmatrix}.$$

We now find the Smith form of $\lambda I - J$ by elementary transformations

$$\lambda I - J = \begin{bmatrix} \lambda - 1 & 0 & 0 \\ 0 & \lambda - 2 & 0 \\ 0 & 0 & \lambda - 2 \end{bmatrix}$$

$$\sim \begin{bmatrix} \lambda - 1 & 0 & 0 \\ 0 & \lambda - 2 & 0 \\ \lambda - 2 & 0 & \lambda - 2 \end{bmatrix} \quad C_1 + C_3$$

$$\sim \begin{bmatrix} 1 & 0 & 2 - \lambda \\ 0 & \lambda - 2 & 0 \\ \lambda - 2 & 0 & \lambda - 2 \end{bmatrix} \quad R_1 - R_3$$

$$\sim \begin{bmatrix} 1 & 0 & 0 \\ 0 & \lambda - 2 & 0 \\ \lambda - 2 & 0 & (\lambda - 2)(\lambda - 1) \end{bmatrix} \quad C_3 - (2 - \lambda)C_1$$

$$\sim \begin{bmatrix} 1 & 0 & 0 \\ 0 & \lambda - 2 & 0 \\ 0 & 0 & (\lambda - 2)(\lambda - 1) \end{bmatrix} \quad R_3 - (\lambda - 2)R_1.$$

Now apply the following elementary column operations to the unit matrix

$$(C_1 \leftrightarrow C_3)(C_3 + \lambda C_2)(C_2 \leftrightarrow C_3)(C_3 - C_1)(C_2 - C_3)[(-1)C_1](C_2 \leftrightarrow C_3) \times [C_3 - (\lambda - 1)C_2][C_3 + (2 - \lambda)C_1](C_1 - C_3)$$

giving

$$\begin{bmatrix} 1 & 0 & 0 \\ 0 & 1 & 0 \\ 0 & 0 & 1 \end{bmatrix} \to \begin{bmatrix} 0 & 0 & 1 \\ 0 & 1 & 0 \\ 1 & 0 & 0 \end{bmatrix} \to \begin{bmatrix} 0 & 0 & 1 \\ 0 & 1 & \lambda \\ 1 & 0 & 0 \end{bmatrix} \to \begin{bmatrix} 0 & 1 & 0 \\ 0 & \lambda & 1 \\ 1 & 0 & 0 \end{bmatrix} \to$$

$$\to \begin{bmatrix} 0 & 1 & 0 \\ 0 & \lambda & 1 \\ 1 & 0 & -1 \end{bmatrix} \to \begin{bmatrix} 0 & 1 & 0 \\ 0 & \lambda - 1 & 1 \\ 1 & 1 & -1 \end{bmatrix} \to \begin{bmatrix} 0 & 1 & 0 \\ 0 & \lambda - 1 & 1 \\ -1 & 1 & -1 \end{bmatrix} \to$$

$$\to \begin{bmatrix} 0 & 0 & 1 \\ 0 & 1 & \lambda - 1 \\ -1 & -1 & 1 \end{bmatrix} \to \begin{bmatrix} 0 & 0 & 1 \\ 0 & 1 & 0 \\ -1 & -1 & \lambda \end{bmatrix} \to \begin{bmatrix} 0 & 0 & 1 \\ 0 & 1 & 0 \\ -1 & -1 & 2\lambda - 2 \end{bmatrix} \to$$

$$\to \begin{bmatrix} -1 & 0 & 1 \\ 1 - \lambda & 1 & 0 \\ 1 - 2\lambda & -1 & 2\lambda - 2 \end{bmatrix}.$$

This last matrix can be written as the polynomial matrix

$$\begin{bmatrix} 0 & 0 & 0 \\ -1 & 0 & 0 \\ -2 & 0 & 2 \end{bmatrix} \lambda + \begin{bmatrix} -1 & 0 & 1 \\ 1 & 1 & 0 \\ 1 & -1 & -2 \end{bmatrix}$$

and substituting J for λ gives for the transforming matrix T,

$$T = \begin{bmatrix} 0 & 0 & 0 \\ -1 & 0 & 0 \\ -2 & 0 & 2 \end{bmatrix} \begin{bmatrix} 1 & 0 & 0 \\ 0 & 2 & 0 \\ 0 & 0 & 2 \end{bmatrix} + \begin{bmatrix} -1 & 0 & 1 \\ 1 & 1 & 0 \\ 1 & -1 & -2 \end{bmatrix} = \begin{bmatrix} -1 & 0 & 1 \\ 0 & 1 & 0 \\ -1 & -1 & 2 \end{bmatrix}.$$

As a check on the computation we have

$$AT = \begin{bmatrix} 0 & 1 & 1 \\ 0 & 2 & 0 \\ -2 & 1 & 3 \end{bmatrix} \begin{bmatrix} -1 & 0 & 1 \\ 0 & 1 & 0 \\ -1 & -1 & 2 \end{bmatrix} = \begin{bmatrix} -1 & 0 & 2 \\ 0 & 2 & 0 \\ -1 & -2 & 4 \end{bmatrix}$$

$$TJ = \begin{bmatrix} -1 & 0 & 1 \\ 0 & 1 & 0 \\ -1 & -1 & 2 \end{bmatrix} \begin{bmatrix} 1 & 0 & 0 \\ 0 & 2 & 0 \\ 0 & 0 & 2 \end{bmatrix} = \begin{bmatrix} -1 & 0 & 2 \\ 0 & 2 & 0 \\ -1 & -2 & 4 \end{bmatrix}.$$

Also $|T| = -1$ so that T is non-singular.

In this particular example the Jordan form has reduced to a diagonal matrix and so A is similar to this diagonal matrix. This, of course, means that the columns of T are characteristic vectors of A and this is easily verified numerically. Since T is non-singular the characteristic vectors of A are linearly independent and A is not defective (Section 2–4). Since, however, A has a root of multiplicity 2 it follows that any linear combination of the characteristic vectors (0, 1, −1), (1, 0, 2) associated with the eigenvalue 2 is also a characteristic vector of A [Wilkinson (1965)] associated with the same characteristic root. Thus 3(0, 1, −1) + 2(1, 0, 2) = (2, 3, 1) and

$$\begin{bmatrix} 0 & 1 & 1 \\ 0 & 2 & 0 \\ -2 & 1 & 3 \end{bmatrix} \begin{bmatrix} 2 \\ 3 \\ 1 \end{bmatrix} = 2 \begin{bmatrix} 2 \\ 3 \\ 1 \end{bmatrix}.$$

Gantmacher (1959, p. 164) gives a second technique for constructing the transformation matrix T for use when the elementary divisors of A are known in advance. The method is rather more complex and we refer the reader to the above text for details.

3–2 The companion form

A matrix C of the form

$$C = \begin{bmatrix} 0 & 1 & 0 & 0 & \cdots & 0 \\ 0 & 0 & 1 & 0 & \cdots & 0 \\ 0 & 0 & 0 & 1 & \cdots & 0 \\ \cdot & \cdot & \cdot & \cdot & \cdots & \cdot \\ \cdot & \cdot & \cdot & \cdot & \cdots & \cdot \\ 0 & 0 & 0 & \cdot & \cdots & 1 \\ -c_n & -c_{n-1} & -c_{n-2} & -c_{n-3} & \cdots & -c_1 \end{bmatrix} \tag{3–2–1}$$

is called a *companion matrix* and is associated with the polynomial

$$\lambda^n + c_1\lambda^{n-1} + \cdots + c_n \tag{3–2–2}$$

since (3–2–2) is, in fact, the characteristic polynomial of (3–2–1). This is easily seen by direct expansion of the characteristic determinant of (3–2–1) (by first column).

(We restrict our use of the name 'companion matrix' to (3–2–1) but the name is also used for similar forms with the c_i in the first row, first column or last column. MacDuffee seems to be the first to have called (3–2–1) the companion form in 1933 [see MacDuffee (1956)].)

Exercise 3–2–1

Prove that the polynomial (3–2–2) is also the minimal polynomial of C so that any companion matrix is non-derogatory.

Since the similarity invariants of any non-derogatory matrix A consist of its minimal polynomial (or characteristic polynomial) and $(n - 1)$ units (Theorem 2–5–9) it follows that if (3–2–2) is the characteristic polynomial of A then A is similar to C (Theorem 2–5–6). We now discuss the transforming matrix involved in the similarity transformation of an arbitrary non-derogatory $n \times n$ matrix A into companion form C.

Since A is non-derogatory there is only a single Jordan block associated with each distinct eigenvalue λ_i of A (Section 2–4). It follows that A is similar to $J = \text{diag}\,(J_1, \ldots, J_r, \ldots, J_p)$ and this is in turn similar to C. In the last section it was shown how to determine T such that

$$T^{-1}AT = J. \tag{3–2–3}$$

A matrix H is now defined such that

$$HJH^{-1} = C \tag{3–2–4}$$

and then from (3–2–3) and (3–2–4) it follows that

$$K^{-1}AK = C$$

where

$$K = TH^{-1}.$$

Suppose that J_r is of order m* and corresponds to λ_r and let H_r be defined so that

$$H_rJ_r = \begin{bmatrix} 1 & 0 & \cdots & 0 \\ \lambda_r & 1 & \cdots & 0 \\ \cdot & \cdot & \cdots & \cdot \\ \cdot & \cdot & \cdots & \cdot \\ \cdot & \cdot & \cdots & \cdot \\ \lambda_r^{n-2} & \binom{n-2}{1}\lambda_r^{n-3} & \cdots & \binom{n-2}{m-1}\lambda_r^{n-m-1} \\ \lambda_r^{n-1} & \binom{n-1}{1}\lambda_r^{n-2} & \cdots & \binom{n-1}{m-1}\lambda_r^{n-m} \end{bmatrix} J_r$$

$$= \begin{bmatrix} \lambda_r & 1 & \cdots & 0 \\ \lambda_r^2 & 2\lambda_r & \cdots & 0 \\ \cdot & \cdot & \cdots & \cdot \\ \cdot & \cdot & \cdots & \cdot \\ \cdot & \cdot & \cdots & \cdot \\ \lambda_r^{n-1} & \binom{n-1}{1}\lambda_r^{n-2} & \cdots & \binom{n-1}{m-1}\lambda_r^{n-m} \\ \lambda_r^{n} & \binom{n}{1}\lambda_r^{n-1} & \cdots & \binom{n}{m-1}\lambda_r^{n-m+1} \end{bmatrix}.$$

But CH_r is given by

$$\begin{bmatrix} 0 & 1 & 0 & 0 & \cdots & 0 \\ 0 & 0 & 1 & 0 & \cdots & 0 \\ 0 & 0 & 0 & 1 & \cdots & 0 \\ \cdot & \cdot & \cdot & \cdot & \cdots & \cdot \\ \cdot & \cdot & \cdot & \cdot & \cdots & \cdot \\ \cdot & \cdot & \cdot & \cdot & \cdots & \cdot \\ \cdot & \cdot & \cdot & \cdot & \cdots & \cdot \\ -c_n & -c_{n-1} & -c_{n-2} & \cdots & & c_1 \end{bmatrix} \begin{bmatrix} 1 & 0 & \cdots & 0 \\ \lambda_r & 1 & \cdots & 0 \\ \lambda_r^2 & \lambda_r & \cdots & 0 \\ \cdot & \cdot & \cdots & \cdot \\ \cdot & \cdot & \cdots & \cdot \\ \cdot & \cdot & \cdots & \cdot \\ \lambda_r^{n-2} & \binom{n-2}{1}\lambda_r^{n-3} & \cdots & \binom{n-2}{m-1}\lambda_r^{n-m-1} \\ \lambda_r^{n-1} & \binom{n-1}{1}\lambda_r^{n-2} & \cdots & \binom{n-1}{m-1}\lambda_r^{n-m} \end{bmatrix}$$

$$= H_rJ_r$$

* There is slight difference in notation here from that in Section 2–4.

since λ_r is a root of the characteristic polynomial with multiplicity m. (This means that the first $m - 1$ derivatives of (3–2–2) vanish so, e.g.,

$$-c_{n-1} - c_{n-2}\,\lambda_r \cdots - c_1\binom{n-1}{1}\lambda_r^{n-2} = \binom{n}{1}\lambda_r^{n-1}.)$$

It follows that the matrix $H = (H_1, \ldots, H_p)$ can be used as a similarity transformation from J into the companion form C.

In the special case when $J = \text{diag}\,(\lambda_1, \lambda_2, \ldots, \lambda_n)$ then, of course, H takes the particularly simple form

$$\begin{bmatrix} 1 & 1 & \cdots & 1 \\ \cdot & \cdot & \cdots & \cdot \\ \cdot & \cdot & \cdots & \cdot \\ \cdot & \cdot & \cdots & \cdot \\ \lambda_1^{n-2} & \lambda_2^{n-2} & \cdots & \lambda_n^{n-2} \\ \lambda_1^{n-1} & \lambda_2^{n-1} & \cdots & \lambda_n^{n-1} \end{bmatrix}.$$

Example 3–2–1

The 3×3 matrix $\begin{bmatrix} 1 & 0 & 0 \\ 0 & 2 & 0 \\ 0 & 0 & 3 \end{bmatrix}$

has characteristic equation $(\lambda - 1)(\lambda - 2)(\lambda - 3) = 0$ or $\lambda^3 - 6\lambda^2 + 11\lambda - 6 = 0$ so that its companion matrix is

$$\begin{bmatrix} 0 & 1 & 0 \\ 0 & 0 & 1 \\ 6 & -11 & 6 \end{bmatrix}.$$

Hence we have $\begin{bmatrix} 0 & 1 & 0 \\ 0 & 0 & 1 \\ 6 & -11 & 6 \end{bmatrix} \begin{bmatrix} 1 & 1 & 1 \\ 1 & 2 & 3 \\ 1 & 4 & 9 \end{bmatrix} = \begin{bmatrix} 1 & 2 & 3 \\ 1 & 4 & 9 \\ 1 & 8 & 27 \end{bmatrix}$

$$= \begin{bmatrix} 1 & 1 & 1 \\ 1 & 2 & 3 \\ 1 & 4 & 9 \end{bmatrix} \begin{bmatrix} 1 & 0 & 0 \\ 0 & 2 & 0 \\ 0 & 0 & 3 \end{bmatrix}.$$

It is also possible to transform an arbitrary non-derogatory matrix A directly (i.e., without first going to the Jordan form) into companion form by the following procedure [Tuel (1966), Wonham and Johnson (1964)]. Let K be the required transforming matrix, then

$$KAK^{-1} = C \qquad (3\text{–}2\text{–}5)$$

and K is given by

$$K = \begin{bmatrix} f \\ fA \\ fA^2 \\ \cdot \\ \cdot \\ \cdot \\ fA^{n-1} \end{bmatrix} \tag{3–2–6}$$

where f is any row vector which makes K non-singular. From Eqns (3–2–5) and (3–2–6) it is apparent that

$$c_{n-r+1} = -fA^n g_r, \; r = 1, 2, \ldots, n$$

where g_r represents the rth column of K^{-1}. For details of a computer programme for carrying out the transformation (3–2–5) Barnett (1967) can be consulted. [See also Theorem 2–2–2 for a direct method of calculating the c_i in (3–2–1) and Ramaswami and Ramar (1968) for an efficient computing algorithm and further references.]

This result arises from the more general problem of carrying out a transformation $y = Kx$ which will reduce the differential system

$$\dot{x} = Ax + fu$$

into the form

$$\dot{y} = Cy + gu$$

where C is in companion form and the row vector g' is $(0, 0, \ldots, 1)$ [Wonham and Johnson (1964)]. The result follows from partitioning K into its rows $K' = [k'_1, k'_2, \ldots, k'_n]$ and using the equations $KA = CK$ and $Kf = g$.

3–3 The Frobenius canonical form

If the $n \times n$ matrix A is derogatory then it is not possible to reduce it to its associated companion form by a similarity transformation. This follows from the fact that the matrix $(C - \lambda I)$ has at least rank $n - 1$ no matter what the value of λ, or from Theorem 2–5–6. We can, however, state the following theorem for matrix A derogatory or otherwise.

Theorem 3–3–1

A matrix A can be transformed into the direct sum of a set of matrices which are each of the same form as the companion matrix by a similarity transformation, i.e., there exists a non-singular matrix H such that

$$H^{-1}AH = \text{diag} \;\; (F_1, F_2, \ldots, F_r) \tag{3–3–1}$$

where
$$F_j = \begin{bmatrix} 0 & 1 & 0 & \cdots & 0 \\ 0 & 0 & 1 & \cdots & 0 \\ \cdot & \cdot & \cdot & \cdots & \cdot \\ \cdot & \cdot & \cdot & \cdots & \cdot \\ \cdot & \cdot & \cdot & \cdots & \cdot \\ -f_{s-1} & -f_{s-2} & \cdot & \cdots & f_0 \end{bmatrix} \qquad j = 1, 2, \ldots, r.$$

The matrix on the right side of (3–3–1) is known as the *Frobenius canonical* form of A. The characteristic polynomial of each of the F_j can be shown to divide that of all the preceding matrices.

The Frobenius form F of a matrix A can be easily obtained from its Jordan form J in the following way. The Jordan blocks of highest order in each characteristic root are combined to form a single matrix in the upper left diagonal position of J, then the Jordan blocks of next highest order in each characteristic root are combined into a single matrix in the next diagonal position and so on. When combined in this way the resulting matrices are all non-derogatory. Each of the non-derogatory matrices can then be transformed into its companion form by a similarity transformation; the result is F.

3–4 Schwarz form

A matrix W is said to be in *Schwarz form* if

$$W = \begin{bmatrix} 0 & 1 & 0 & \cdot & \cdots & \cdot & 0 \\ -w_n & 0 & 1 & \cdot & \cdots & \cdot & 0 \\ 0 & -w_{n-1} & 0 & 1 & \cdots & \cdot & 0 \\ \cdot & \cdot & \cdot & \cdot & \cdots & \cdot & \cdot \\ \cdot & \cdot & \cdot & \cdot & \cdots & 0 & 1 \\ \cdot & \cdot & \cdot & \cdot & \cdots & -w_2 & -w_1 \end{bmatrix}. \qquad (3\text{–}4\text{–}1)$$

It was shown by Schwarz (1956) that such a matrix W could be found similar to any real non-derogatory $n \times n$ matrix A and that the number of characteristic roots of A with negative real parts is given by the number of positive terms in the sequence

$$w_1,\ w_1w_2,\ w_1w_2w_3,\ \ldots,\ w_1w_2 \ldots w_n$$

assuming none of the w_i are zero.

Schwarz also showed how W could be obtained from A using a finite sequence of elementary transformations and considered both real and complex matrices A. The form of W is, of course, different for complex A. (See also Theorem 3–5–2)

We now give a method due to Loo (1968) [see also Butchart (1965)] for transforming a matrix in companion form C into Schwarz form W by a similarity transformation

$$W = TCT^{-1} \tag{3–4–2}$$

where the transforming matrix T is lower triangular.

The elements of T are obtained directly by equating appropriate elements in the equation $WT = TC$ and are given by the following rules:

1. $t_{11} = t_{22} = \cdots = t_{nn} = 1$.
2. All the odd subdiagonals of T up to the qth have all elements zero where

$$q = \frac{n}{2} - 1, \quad n \text{ even}$$

$$q = \frac{n-1}{2}, \quad n \text{ odd}$$

3. Other non-zero elements of T are obtained from

$$t_{s+2l,s} = d'_{n-s-2l+2,\,l+1}$$

$$\text{for} \quad l = 1, 2, \ldots, q$$

$$s = 1, 2, \ldots, n - 2l.$$

Here q is as given above and d'_{ij} represents the ratio of the ijth to the $i1$th element of the Routh array for the polynomial (3–2–2).

For completeness we give below the Routh array for the characteristic polynomial of C

$$d_{11} = 1$$

$$d_{1j} = c_{2(j-1)},\ 2 \leqslant j \leqslant m + 1, \quad d_{1j} = 0, j > m + 1$$

$$d_{2j} = (1 - \delta_{j,\frac{1}{2}n+1})\, c_{2j-1},\ 1 \leqslant j \leqslant m+1, \quad d_{2j} = 0, j > m+1$$

$$d_{3,m+1} = d_{4,m+1} = \cdots = d_{n+1,m+1} = 0$$

and for n even, $m = \frac{n}{2}$, for n odd, $m = \frac{n-1}{2}$.

The remaining elements are given by

$$d'_{i-2,j+1} = (1 - \delta_{i,n-2j+3})d'_{i-1,j+1} + \frac{d'_{ji}d_{i1}}{d_{i-2,1}}$$

for $j = 1, 2, \ldots, q$, $i = 4, 5, \ldots, n - 2j + 3$ where q is again given above.

Exercise 3–4–1 (Loo, 1968)

As a result of equating elements of the last rows of WT and TC show that

$$c_{n-s+1} = \frac{1}{t_{nn}}\,[t_{ns}w_1 + t_{n-1,s}w_2 + (1 - \delta_{s1})t_{n,s-1}]$$

for $s = 1, 2, \ldots, n$.

This shows that the c_i can be calculated from w_1, w_2 and the elements of the last two rows of T.

For a method of finding a transforming matrix T in terms of the elements of W rather than those of the Routh array we refer the reader to Power (1967). For details of a direct algorithm for transforming from the Schwarz to the Jordan form and a discussion of the interrelationships between the companion, Jordan and Schwarz forms the paper by Sarma, Pai, and Viswanathan (1968*a*) should be consulted. Finally, we mention the paper by Sarma *et al.* (1968*b*) for a direct method of obtaining the Schwarz form of an arbitrary non-derogatory matrix. This technique is based on the same idea as that of Tuel (1966).

3–5 Jacobi matrices

A tridiagonal matrix

$$G = \begin{bmatrix} b_1 & c_1 & 0 & \cdots & 0 & 0 \\ a_1 & b_2 & c_2 & \cdots & 0 & 0 \\ 0 & a_2 & b_3 & \cdots & 0 & 0 \\ \cdot & \cdot & \cdot & \cdots & \cdot & \cdot \\ \cdot & \cdot & \cdot & \cdots & \cdot & \cdot \\ \cdot & \cdot & \cdot & \cdots & \cdot & \cdot \\ 0 & 0 & 0 & \cdots & a_{n-1} & b_n \end{bmatrix} \tag{3–5–1}$$

is said to be a Jacobi matrix if

$$a_i c_i > 0 \text{ for } i = 1, 2, \ldots, n-1. \tag{3–5–2}$$

If the elements of G are real then:

Theorem 3–5–1

The characteristic roots of G are all real.

Proof. Let G be a Jacobi matrix of order n and D the diagonal matrix diag $(d_1, d_2, \ldots, d_n)$. Letting

$$R = DGD^{-1} \tag{3–5–3}$$

a direct calculation shows that

$$\begin{aligned} r_{ij} &= b_j, \qquad && i = j \\ r_{ij} &= 0, && |i-j| \geqslant 2 \\ r_{ij} &= \frac{d_i}{d_{i+1}} c_i \\ r_{ji} &= \frac{d_{i+1}}{d_i} a_i \end{aligned} \left.\vphantom{\frac{d_i}{d_{i+1}}\frac{d_i}{d_{i+1}}}\right\} \; j = i+1$$

for $i, j = 1, 2, \ldots, n$. Because of (3–5–2) it follows that the elements d_i of D can be selected in such a way that

$$\frac{d_i}{d_{i+1}} = \left(\frac{a_i}{c_i}\right)^{1/2}$$

so that R is symmetric. Hence G is similar to a real symmetric matrix and therefore has real characteristic roots (Theorem 2–3–4).

Exercise 3–5–1

By expanding the characteristic determinant in terms of its last row show that the characteristic polynomial $\phi_n(\lambda)$ of an $n \times n$ Jacobi matrix can be obtained from the recurrence formula

$$\phi_n(\lambda) = (\lambda - b_n)\phi_{n-1} - c_{n-1}\,a_{n-1}\,\phi_{n-2}, \quad n > 1$$

$$\phi_1(\lambda) = \lambda - b_1$$

$$\phi_0(\lambda) = 1.$$

This recurrence formula is, of course, true for any tridiagonal determinant.

Exercise 3–5–2

Prove that the inverse of an $n \times n$ Schwarz matrix W is given by

$$W^{-1} = \begin{bmatrix}
\dfrac{-w_2w_4\cdots w_{n-1}}{w_1w_3\cdots w_n} & \dfrac{-1}{w_1} & \dfrac{-w_4w_6\cdots w_{n-1}}{w_1w_3\cdots w_n} & \dfrac{-1}{w_1w_3} & \cdots & \dfrac{-1}{w_1w_3\cdots w_n} \\
1 & 0 & 0 & 0 & \cdots & 0 \\
\dfrac{-w_2w_4\cdots w_{n-1}}{w_3w_5\cdots w_n} & 0 & \dfrac{-w_4w_6\cdots w_{n-1}}{w_3w_5\cdots w_n} & \dfrac{-1}{w_3} & \cdots & \dfrac{-1}{w_3w_5\cdots w_n} \\
w_2 & 0 & 1 & 0 & \cdots & 0 \\
\dfrac{-w_2w_4\cdots w_{n-1}}{w_5w_7\cdots w_n} & 0 & \dfrac{-w_4w_6\cdots w_{n-1}}{w_5w_7\cdots w_{n-1}} & 0 & \cdots & \dfrac{-1}{w_5w_7\cdots w_n} \\
w_2w_4 & 0 & w_4 & 0 & \cdots & 0 \\
\cdot & \cdot & \cdot & \cdot & \cdots & \cdot \\
\cdot & \cdot & \cdot & \cdot & \cdots & \cdot \\
\cdot & \cdot & \cdot & \cdot & \cdots & \cdot \\
w_2w_4\cdots w_{n-3} & 0 & w_4\cdots w_{n-3} & 0 & \cdots & 0 \\
\dfrac{-w_2w_4\cdots w_{n-1}}{w_n} & 0 & \dfrac{-w_4w_6\cdots w_{n-1}}{w_n} & 0 & \cdots & \dfrac{-1}{w_n}
\end{bmatrix}$$

if n is odd, and

$$W^{-1} = \begin{bmatrix} \dfrac{-w_2w_4\cdots w_n}{w_1w_3\cdots w_{n-1}} & \dfrac{-1}{w_1} & \dfrac{-w_4\cdots w_n}{w_1\cdots w_{n-1}} & \dfrac{-1}{w_1w_3} & \cdots & \dfrac{-1}{w_1\cdots w_{n-1}} \\ 1 & 0 & 0 & 0 & \cdots & 0 \\ \cdot & \cdot & \cdot & \cdot & \cdots & \cdot \\ \cdot & \cdot & \cdot & \cdot & \cdots & \cdot \\ \cdot & \cdot & \cdot & \cdot & \cdots & \cdot \\ \dfrac{-w_2\cdots w_n}{w_{n-1}} & 0 & \dfrac{-w_4\cdots w_n}{w_{n-1}} & 0 & \cdots & \dfrac{-1}{w_{n-1}} \\ w_2w_4\cdots w_{n-2} & 0 & w_4\cdots w_{n-2} & 0 & \cdots & 0 \end{bmatrix}$$

if n is even.

To conclude this chapter we give a useful result concerning the characteristic roots of complex tridiagonal matrices.

Theorem 3–5–2 (*Schwarz*)
Let the diagonal elements of a tridiagonal matrix G be complex with $Re\,(b_1) \neq 0, b_2, b_3, \ldots, b_n$ pure imaginary, $a_i = -1, i = 1, 2, \ldots, n-1$, and $c_j, j = 1, 2, \ldots, n-1$, non-zero real numbers. Then the number of characteristic roots of G with positive real parts is equal to the number of positive elements of the sequence $\{c_0c_1 \ldots c_{r-1}\}, r = 1, \ldots, n$ where $c_0 = Re\,(b_1)$. It follows that if all the c_j are positive then all the characteristic roots have positive real parts.

Discussion and further reading

In this chapter a number of special forms of matrix which are especially useful in stability analysis have been discussed. We have also given some account of their relationships to one another and to similarity transformations between them. For more information on the classical canonical forms and the Jacobi matrix we refer the reader to Gantmacher (1959), Marcus and Minc (1964), and Mishina and Proskuryakov (1965). For fuller details concerning the Schwarz form the papers by Schwarz (1956), Loo (1968), and Sarma *et al.* (1968*a*, *b*) should be consulted.

References

BARNETT, S. (1967), Solution of some algebraic problems arising in the theory of stability and sensitivity of systems, with particular reference to the Liapunov matrix equation, Ph.D. Thesis, Loughborough University of Technology.

BUTCHART, R. L. (1965), An explicit solution to the Fokker-Planck equation for an ordinary differential equation. *Int. J. Control*, **1**, 201–208.

GANTMACHER, F. R. (1959), *The theory of matrices* Chelsea Publishing Co., New York.

Loo, S. G. (1968), A simplified proof of a transformation matrix relating the companion matrix and the Schwarz matrix, *I.E.E.E. Trans. Aut. Control*, **AC–13,** 309–310.

MacDuffee, C. C. (1956), *The theory of matrices*, Chelsea Publishing Co., New York.

Marcus, M. and H. Minc (1964), *A survey of matrix theory and matrix inequalities*, Allyn and Bacon, Boston.

Mishina, A. P. and I. V. Proskuryakov (1965), *Higher Algebra*, Pergamon Press, Oxford.

Power, H. M. (1967), The companion matrix and Lyapunov functions for linear, multivariable, time-invariant systems, *J. Franklin Inst.*, **283,** 214–234.

Ramaswami, B. and K. Ramar (1968), Transformation to the phase-variable canonical form, *I.E.E.E. Trans. Aut. Control*, **AC–13,** 746–747.

Sarma, I. G., M. A. Pai and R. Viswanathan (1968*a*), A note on the transformation from Schwarz to Jordan canonical form, *I.E.E.E. Trans. Aut. Control*, **AC–13,** 310–311.

Sarma, I. G., M. A. Pai and R. Viswanathan (1968*b*), On the transformation to Schwarz canonical form, *I.E.E.E. Trans. Aut. Control*, **AC–13**, 311–312.

Schwarz, H. R. (1956), Ein Verfahren zur Stabilitätsfrage bei Matrizen-Eigenwerteprobleme, *Z. Angew. Math. Phys.*, **7,** 473–500.

Tuel, W. G. jr (1966), On the transformation to (phase-variable) canonical form, *I.E.E.E. Trans. Aut. Control*, **AC–11**, 607. (See also subsequent correspondence in this issue.)

Wilkinson, J. H. (1965), *The algebraic eigenvalue problem*, Oxford University Press, London.

Wonham, W. M. and C. D. Johnson (1964), Optimal bang-bang control with quadratic performance index, *Trans. A.S.M.E.*, Series D, **86,** 107–115.

4

Functions of a matrix

4–1 Matrix and vector series

Any sequence $\{x_k\}$ of vectors $x_k = (\xi_{1k}, \xi_{2k}, \ldots, \xi_{nk})$ is said to *tend to the limit* $x = (\xi_1, \xi_2, \ldots, \xi_n)$ if

$$\underset{k \to \infty}{Lt} \ \xi_{ik} = \xi_i, \quad i = 1, 2, \ldots, n$$

or, in words, the components of the limiting vector are the limiting values of the corresponding components of the sequence. In a similar manner a sequence $\{A_k\}$ of $n \times n$ matrices $A_k = (a_{ij}^k)$ is said to tend to the limit $A = (a_{ij})$ if

$$\underset{k \to \infty}{Lt} \ a_{ij}^k = a_{ij}, \quad i, j = 1, 2, \ldots, n.$$

It is a natural extension of the corresponding result for a scalar series to say that the infinite vector series $\sum_{k=1}^{\infty} x_k$ *converges to the sum* x if the sequence of partial sums

$$X_n = \sum_{k=0}^{n} x_k$$

tends to the limit x as $n \to \infty$. Similarly, the infinite matrix series $\sum_{k=0}^{\infty} A_k$ converges to the sum A provided that the sequence of partial sums

$$Y_n = \sum_{k=0}^{n} A_k$$

tends to A as $n \to \infty$.

We can thus define the *derivative* of a matrix

$$A(t) = [a_{ij}(t)]$$

where t is a parameter, in the following way:

$$\frac{dA}{dt} = \underset{\delta t \to 0}{Lt} \frac{A(t + \delta t) - A(t)}{\delta t}. \tag{4–1–1}$$

This, of course, simply means that

$$\frac{dA}{dt} = \left(\frac{da_{ij}(t)}{dt}\right) \tag{4–1–2}$$

and a matrix $A(t)$ is differentiable if its elements are differentiable functions of the parameter t.

Exercise 4–1–1

Prove the following results concerning the differentiation of matrices:

(i) $\frac{d}{dt}[A_1(t) + A_2(t)] = \frac{dA_1(t)}{dt} + \frac{dA_2(t)}{dt}$

(ii) $\frac{d}{dt}[\alpha A(t)] = \alpha \frac{d}{dt} A(t)$ (α a scalar)

(iii) $\frac{d}{dt}[A_1(t)A_2(t)] = A_1(t)\frac{dA_2(t)}{dt} + \frac{dA_1(t)}{dt}A_2(t)$

(Notice that $\frac{d}{dt}[A^2(t)] = A(t)\frac{dA(t)}{dt} + \frac{dA(t)}{dt}A(t)$

$$= 2A(t)\frac{dA(t)}{dt}$$

only if $\frac{dA(t)}{dt}$ and $A(t)$ commute.)

(iv) $\frac{d}{dt}[A^{-1}(t)] = -A^{-1}(t)\frac{dA(t)}{dt}A^{-1}(t)$.

Exercise 4–1–2

Prove that a necessary and sufficient condition for $\underset{n\to\infty}{Lt}\, A_n = A$ is that $\underset{n\to\infty}{Lt}\, \| A_n - A \| = 0$ (for any norm).

Interpret the result for $A = 0$.

Exercise 4–1–3

Show that if

$$\underset{n\to\infty}{Lt} \sum_{i=1}^{n} A_i = A$$

then

$$\underset{n\to\infty}{Lt} \sum_{i=1}^{n} BA_iC = BAC.$$

The matrix series $\sum_{k=0}^{\infty} A_k$ is said to be *absolutely convergent* if $\sum_{k=0}^{\infty} \| A_k \|$ is convergent (any norm can be used: a suitable one being the cubic matrix norm $\| A_k \|_c$ which is consistent with the euclidean vector norm). It is easily seen from the corresponding result for scalar series that any absolutely convergent matrix series is also convergent. Thus if $A_k = (a_{ij}^k)$ then $| a_{ij}^k | \leqslant \| A_k \|_c$ (Section 2–9) and the convergence of $\sum_{k=0}^{\infty} \| A_k \|$ implies that of $\sum_{k=0}^{\infty} | a_{ij}^k |$ and hence that of $\sum_{k=0}^{\infty} a_{ij}^k$.

Exercise 4–1–4
Prove that geometric matrix series $I_n + A + A^2 + \cdots$ is absolutely convergent provided $\| A \| < 1$.

The spectral radius $\rho(A)$ of a matrix A was defined in Section 2–9 as

$$\rho(A) = \max_i \{| \lambda_i(A) |\} \tag{4–1–3}$$

and if $\| A \|$ is any matrix norm Theorem 2–9–2 states that $\rho(A) < \| A \|$.
That the matrix geometric series

$$I + A + A^2 + \cdots + A^n + \cdots \tag{4–1–4}$$

is convergent when $\rho(A) < 1$ can be seen as follows. From $\rho(A) < 1$ we infer immediately that $\det(I - A) \neq 0$ so that A is non-singular. Now let the partial sum

$$S_n = I + A + A^2 + \cdots + A^n$$

so that

$$S_n(I - A) = I - A^{n+1}.$$

The result

$$\underset{n \to \infty}{Lt} \; S_n(I - A) = I$$

now follows from Theorem 2–9–4 and it is seen that the sum of the geometric series (4–1–4) is just $(I - A)^{-1}$.

More general results concerning the convergence of matrix power series are given in the following theorems.

Theorem 4–1–1
The matrix power series

$$f(A) = a_0I + a_1A + a_2A^2 + \cdots \tag{4–1–5}$$

is convergent if the associated (complex) scalar power series

$$f(x) = a_0 + a_1x + a_2x^2 + \cdots \tag{4–1–6}$$

has a radius of convergence $r > \rho(A)$. When (4–1–5) is convergent its sum is defined to be $f(A)$.

Proof. Let $J = S^{-1}AS$ be the Jordan form of A and consider the partial sum

$$\begin{aligned} f_N(A) &= \sum_{i=0}^{N} a_i A^i \\ &= \sum_{i=0}^{N} a_i \,(SJS^{-1})^i \\ &= S[\sum_{i=0}^{N} a_i J^i] \, S^{-1}. \end{aligned} \tag{4–1–7}$$

But $J^n = \text{diag}\,(J_1^n, J_2^n, \ldots, J_s^n)$ where J_k, $k = 1, 2, \ldots, s$ are the Jordan blocks of A and

$$J_k^n = \begin{bmatrix} \lambda^n & \binom{n}{1}\lambda^{n-1} & \binom{n}{2}\lambda^{n-2} & \cdots \\ 0 & \lambda^n & \binom{n}{1}\lambda^{n-1} & \cdots \\ 0 & 0 & \lambda^n & \cdots \\ \cdot & \cdot & \cdot & \cdots \\ \cdot & \cdot & \cdot & \cdots \\ \cdot & \cdot & \cdot & \cdots \end{bmatrix}. \qquad (4\text{–}1\text{–}8)$$

(See Exercise 3–1–1.)

Using (4–1–8) in (4–1–7) it is easily seen that each of the scalar series converges provided $r > \rho(A)$ and hence the matrix series (4–1–5) converges.

In the special case when A has n distinct characteristic roots $J = \operatorname{diag}(\lambda_1, \lambda_2, \ldots, \lambda_n)$ and the partial sum in (4–1–7) takes the particularly simple form

$$f_N(A) = S[\sum_{i=0}^{N} a_i \operatorname{diag}(\lambda_1^i, \lambda_2^i, \ldots, \lambda_n^i)]S^{-1}$$

giving

$$f(A) = S \operatorname{diag}[f(\lambda_1), f(\lambda_2), \ldots, f(\lambda_n)]S^{-1}.$$

A stronger form of Theorem 4–1–1 is the following:

Theorem 4–1–2

A necessary and sufficient condition for convergence of the matrix power series (4–1–5) is that (i) $\rho(A) \leqslant r$ and (ii) for every characteristic root λ_i of A, of multiplicity s, for which $|\lambda_i| = r$, the power series $f^{(s-1)}(\lambda_i)$ is also convergent.

A proof of this theorem can be found in MacDuffee (1946).

Theorem 4–1–3

If the series (4–1–6) is convergent in the whole complex plane then (4–1–5) is absolutely convergent for an arbitrary matrix A.

For example, since the scalar power series for the functions exp x, cos x, and sin x are convergent everywhere in the complex plane it follows that

$$\exp A = I + A + \frac{1}{2!}A^2 + \frac{1}{3!}A^3 + \cdots$$

$$\cos A = I - \frac{1}{2!}A^2 + \frac{1}{4!}A^4 - \cdots$$

$$\sin A = A - \frac{1}{3!}A^3 + \frac{1}{5!}A^5 - \cdots$$

converge for all A. On the other hand $\ln(1 + x)$ converges for all x for which $|x| < 1$ showing that

$$\ln(I + A) = A - \tfrac{1}{2}A^2 + \tfrac{1}{3}A^3 - \cdots$$

converges for all matrices A having $\rho(A) < 1$.

4–2 Functions of a matrix

Given a general function $f(\gamma)$ of some scalar variable γ we now briefly discuss the problem of how to interpret $f(A)$, i.e., the result of replacing γ by an $n \times n$ matrix A and a scalar k by kI_n. Of course, we are already quite familiar with the special case when f is a polynomial function. Thus if $f(\gamma) = \sum_{k=0}^{n} a_k \gamma^k$ then $f(A)$ is the matrix $\sum_{k=0}^{n} a_k A^k = a_n A^n + a_{n-1} A^{n-1} + \cdots + a_0 I_n$.

Exercise 4–2–1

If J is an nth-order Jordan block and $g(\lambda)$ is the polynomial $\sum_{k=0}^{n} a_k \lambda^k$ show that

$$g(J) = \begin{bmatrix} g(\lambda) & \dfrac{g'(\lambda)}{1!} & \dfrac{g''(\lambda)}{2!} & \dfrac{g'''(\lambda)}{3!} & \cdots \\ 0 & g(\lambda) & \dfrac{g'(\lambda)}{1!} & \dfrac{g''(\lambda)}{2!} & \cdots \\ 0 & 0 & g(\lambda) & \dfrac{g'(\lambda)}{1!} & \cdots \\ \cdot & \cdot & \cdot & \cdot & \cdots \\ \cdot & \cdot & \cdot & \cdot & \cdots \\ \cdot & \cdot & \cdot & \cdot & \cdots \end{bmatrix}$$

where the highest order derivative is the $(n - 1)$th. (Use the expression for J^n given in Section 3–1.)

Now let A be any $n \times n$ matrix with minimum polynomial given by

$$m(\lambda) = (\lambda - \lambda_1)^{j_1}(\lambda - \lambda_2)^{j_2} \cdots (\lambda - \lambda_r)^{j_r} \tag{4–2–1}$$

where λ_i, $i = 1, 2, \ldots, r$, are the distinct characteristic roots of A and $p = \sum_{k=1}^{r} j_k$ is the degree of $m(\lambda)$. The p numbers

$$\left.\begin{matrix} f(\lambda_1), & f'(\lambda_1), & \ldots, & f^{(j_1-1)}(\lambda_1) \\ f(\lambda_2), & f'(\lambda_2), & \ldots & f^{(j_2-1)}(\lambda_2) \\ \cdot & \cdot & \cdots & \cdot \\ \cdot & \cdot & \cdots & \cdot \\ \cdot & \cdot & \cdots & \cdot \\ f(\lambda_r), & f'(\lambda_r), & \ldots, & f^{(j_r-1)}(\lambda_r) \end{matrix}\right\} \tag{4–2–2}$$

are called the *values* of $f(\gamma)$ *on the spectrum of* A and if they all exist $f(\gamma)$ is said to be *defined on the spectrum of* A. In the special case when A has n distinct characteristic roots $\lambda_1, \ldots, \lambda_n$ then $f(\gamma)$ is defined on the spectrum of A when the numbers $f(\lambda_1), f(\lambda_2), \ldots, f(\lambda_n)$ exist.

The fact that the minimum polynomial of A is a factor of any annihilating polynomial of A gives the following theorem:

Theorem 4–2–1

Two scalar polynomials $g(\lambda)$ and $h(\lambda)$ give rise to the same matrix on substituting A for λ, i.e., $g(A) = h(A)$, if and only if g and h have the same values on the spectrum of A.

This result leads naturally to defining $f(A)$ for a general function $f(\lambda)$ as $g(A)$ where $g(\lambda)$ is a polynomial having the same values as $f(\lambda)$ on the spectrum of A. Thus, with this definition, all functions $f(\lambda)$ having the same values on the spectrum of A will result in the same matrix $f(A)$.

Now it is well known from numerical analysis [see, e.g., Wendoff (1966)] that given the set of p numbers (4–2–2) it is possible to find a unique polynomial $g(\lambda)$, of degree not exceeding p, such that

$$\left.\begin{array}{llll} g(\lambda_1) = f(\lambda_1), & g'(\lambda_1) = f'(\lambda_1), & \ldots, & g^{(j_1-1)}(\lambda_1) = f^{(j_1-1)}(\lambda_1) \\ g(\lambda_2) = f(\lambda_2), & g'(\lambda_2) = f'(\lambda_2), & \ldots, & g^{(j_2-1)}(\lambda_2) = f^{(j_2-1)}(\lambda_2) \\ \cdots & \cdots & \cdots & \cdots \\ \cdots & \cdots & \cdots & \cdots \\ \cdots & \cdots & \cdots & \cdots \\ g(\lambda_r) = f(\lambda_r), & g'(\lambda_r) = f'(\lambda_r), & \ldots, & g^{(j_r-1)}(\lambda_r) = f^{(j_r-1)}(\lambda_r) \end{array}\right\} \quad (4\text{–}2\text{–}3)$$

and, in fact, we have the following fundamental theorem.

Theorem 4–2–2

Let the $n \times n$ matrix A have minimum polynomial (4–2–1) and let $f(\lambda)$ be defined on the spectrum of A. Then there exists a unique interpolation polynomial $g(\lambda)$ which satisfies (4–2–3) and which is given by

$$g(\lambda) = \sum_{k=1}^{r} [c_{k1} + c_{k2}(\lambda - \lambda_k) + \cdots + c_{kj_k}(\lambda - \lambda_k)^{j_k - 1}] m_k(\lambda) \quad (4\text{–}2\text{–}4)$$

where the coefficients c_{ki} are given by

$$c_{ki} = \frac{1}{(i-1)!} \left\{ \frac{f(\lambda)}{m_k(\lambda)} \right\}^{(i-1)}_{\lambda = \lambda_k} \quad \text{for} \quad \begin{array}{l} i = 1, 2, \ldots, j_k \\ k = 1, 2, \ldots, r \end{array} \quad (4\text{–}2\text{–}5)$$

and

$$m_k(\lambda) = \frac{m(\lambda)}{(\lambda - \lambda_k)^{j_k}}. \quad (4\text{–}2\text{–}6)$$

On the right side of (4–2–5) the superscript $(i - 1)$ denotes the $(i - 1)$th derivative with respect to λ of the expression in braces and the subscript $\lambda = \lambda_k$ means that the expression is evaluated at λ_k. Notice that the coefficients given by (4–2–5) are fixed once the values of f on the spectrum of A are known. The polynomial $g(\lambda)$ in (4–2–4) is known as the Lagrange-Sylvester polynomial and, by definition, $f(A)$ is $g(A)$.

There are two cases in which $g(\lambda)$ takes a particularly simple form. First if A has n distinct characteristic roots $\lambda_1, \lambda_2 \ldots, \lambda_n$, $g(\lambda)$ becomes

$$g(\lambda) = \sum_{k=1}^{n} \left\{ \prod_{i=1}^{n}{}' \left(\frac{\lambda - \lambda_i}{\lambda_k - \lambda_i} \right) \right\} f(\lambda_k) \tag{4–2–7}$$

where the prime on the product sign means that product is over those factors for which $i \neq k$. The second case is when the minimum polynomial $m(\lambda)$ has no repeated roots but the characteristic polynomial has: if in this case

$$m(\lambda) = (\lambda - \lambda_1)(\lambda - \lambda_2) \cdots (\lambda - \lambda_s)$$

then

$$g(\lambda) = \sum_{k=1}^{s} \left\{ \prod_{i=1}^{s}{}' \left(\frac{\lambda - \lambda_i}{\lambda_k - \lambda_i} \right) \right\} f(\lambda_k). \tag{4–2–8}$$

Of course (4–2–7) and (4–2–8) are straightforward Lagrange interpolation polynomials.

To obtain $f(A)$ in any of these cases it is only necessary to replace λ with A in the expression for $g(\lambda)$. For example, the appropriate equation corresponding to (4–2–8) would be

$$\begin{aligned} f(A) &= g(A) \\ &= \sum_{k=1}^{s} \left\{ \prod_{i=1}^{s}{}' \left(\frac{A - \lambda_i I}{\lambda_k - \lambda_i} \right) \right\} f(\lambda_k). \end{aligned} \tag{4–2–9}$$

It is obvious by inspection that the Lagrange–Sylvester polynomial (4–2–4) can be rearranged in such a way that the terms containing $f(\lambda_1), f'(\lambda_1), \ldots,$ $f(\lambda_2), f'(\lambda_2), \ldots,$ etc. all occur together. The expressions multiplying these terms will be polynomials (of degree less than p) in λ which depend solely on the matrix A and not on the values of f on the spectrum of A. This form of $g(\lambda)$ is particularly useful for computational purposes especially where several different functions of the same matrix A are required. For further details of this approach and for other ways of defining functions of a matrix the reader is referred to Gantmacher (1959) and MacDuffee (1956).

We are now in a position to bring out the natural relationship between the definition of a matrix function in terms of a polynomial and the concept of the limit of an infinite matrix series. The following definition is needed. Let the $n \times n$ matrix A have minimum polynomial $m(\lambda)$ given by (4–2–1). The scalar series $\sum_{k=0}^{\infty} b_k(\lambda)$ is said to *converge on the spectrum of A to the function $f(\lambda)$* provided $b_k(\lambda)$ is defined on the spectrum of A for each k and

$$\left.\begin{array}{l} f(\lambda_1) = \sum_{k=0}^{\infty} b_k(\lambda_1), f'(\lambda_1) = \sum_{k=0}^{\infty} b_k'(\lambda_1), \ldots, f^{(j_1-1)}(\lambda_1) = \sum_{k=0}^{\infty} b_k^{(j_1-1)}(\lambda_1) \\ f(\lambda_2) = \sum_{k=0}^{\infty} b_k(\lambda_2), f'(\lambda_2) = \sum_{k=0}^{\infty} b_k'(\lambda_2), \ldots, f^{(j_2-1)}(\lambda_2) = \sum_{k=0}^{\infty} b_k^{(j_2-1)}(\lambda_2) \\ \cdots\cdots\cdots\cdots\cdots\cdots\cdots\cdots\cdots\cdots \\ f(\lambda_r) = \sum_{k=0}^{\infty} b_k(\lambda_r), f'(\lambda_r) = \sum_{k=0}^{\infty} b_k'(\lambda_r), \ldots, f^{(j_r-1)}(\lambda_r) = \sum_{k=0}^{\infty} b_k^{(j_r-1)}(\lambda_r) \end{array}\right\}$$

The required link is now provided by:

Theorem 4–2–3
A necessary and sufficient condition for the convergence of the matrix series $\sum_{k=0}^{\infty} b_k(A)$ is that $\sum_{k=0}^{\infty} b_k(\lambda)$ should converge on the spectrum of A. Also

$$f(\lambda) = \sum_{k=0}^{\infty} b_k(\lambda) \Leftrightarrow f(A) = \sum_{k=0}^{\infty} b_k(A)$$

where convergence on the spectrum of A is understood in the equation on the left.

The connection between the above theorem and Theorem 4–1–1 should be immediately apparent.

To conclude this section we give a theorem which allows many of the relationships between scalar functions to be extended to the corresponding relationships between matrix functions.

Theorem 4–2–4
Let $f_i(\lambda)$, $i = 1, 2, \ldots, n$, be functions defined on the spectrum of an $n \times n$ matrix A. If any polynomial function of n variables $p(\lambda) \equiv p\{f_i(\lambda)\}$ vanishes on the spectrum of A then $p\{f_i(A)\} = 0$.

Directly from Theorem 4–2–4 it follows that $e^A \,.\, e^{-A} = I_n$, $\cos^2 A + \sin^2 A = I_n$, $\sin 2A = 2 \sin A \cos A$ and so on. Note that these results only concern a single matrix A and say nothing about the corresponding results when two different matrices A and B are involved. It can be said, however, that the results also hold in the latter case provided that the matrices A and B commute. (This is a consequence of the fact that matrices considered as a ring are not commutative.) Thus provided A and B commute $e^A \,.\, e^B = e^{A+B}$, and $\sin(A + B) = \sin A \cos B + \cos A \sin B$, etc.

Exercise 4–2–2
Prove directly from the power series definition of exp A that

$$\exp A \exp B = \exp (A + B)$$

provided A and B commute.

Exercise 4–2–3
Prove that for any positive integer n

$$(A + B)^n = \sum_{k=0}^{n} \binom{n}{k} A^{n-k} B^k$$

provided A and B commute.

Exercise 4–2–4

Prove that the derivative of e^{At} is Ae^{At} and hence that the solution of the matrix equation $\frac{dx}{dt} = Ax$ is $e^{At}x_0$ where $x' = [x_1(t), x_2(t), \ldots, x_n(t)]$ and $x(0)$ equals x_0.

Exercise 4–2–5

Show that e^{At} is non-singular for an arbitrary matrix A and any real number t.

Discussion and further reading

Functions of a matrix have been defined and their calculation in terms of the Lagrange–Sylvester interpolation polynomial discussed. Generally speaking relations which are valid between scalar functions are true for the corresponding matrix functions provided that the matrices involved commute. We have also shown briefly how the concepts of convergence, differentiability, etc., can be carried over to sequences and series of matrices.

For further details on matrix analysis the reader is referred to Pease (1965), Gantmacher (1959), Mirsky (1963), and MacDuffee (1956).

References

GANTMACHER, F. R. (1959), *The theory of matrices*, Vol. I, Chelsea Publishing Co., New York.

MACDUFFEE, C. C. (1956), *The theory of matrices*, Chelsea Publishing Co., New York.

MIRSKY, L. (1963), *An introduction to linear algebra*, Oxford University Press, London.

PEASE, M. C. (1965), *Methods of matrix algebra*, Academic Press, New York.

WENDROFF, B. (1966), *Theoretical numerical analysis*, Academic Press, New York.

5

Liapunov stability theory

5–1 Definitions of stability

Consider the set of first-order, ordinary differential equations of the form

$$\frac{dx_i}{dt} = f_i(x_1, x_2, \ldots, x_n, t), \quad i = 1, 2, \ldots, n. \tag{5–1–1}$$

A convenient shorthand method of writing (5–1–1) is in matrix notation

$$\frac{dx}{dt} = f(x, t) \tag{5–1–2}$$

with $x = (x_1, \ldots, x_n)'$ and $f = (f_1, \ldots, f_n)'$. Notice that the general nth-order differential equation

$$\frac{d^n x}{dt^n} = F\left(\frac{d^{n-1}x}{dt^{n-1}}, \frac{d^{n-2}x}{dt^{n-2}}, \ldots, x, t\right) \tag{5–1–3}$$

can be put into this form by making the substitutions

$$\left.\begin{aligned} x_1 &= x \\ x_2 &= \frac{dx}{dt} \\ x_3 &= \frac{d^2x}{dt^2} \\ &\cdots\cdot \\ x_n &= \frac{d^{n-1}x}{dt^{n-1}}. \end{aligned}\right\} \tag{5–1–4}$$

When the time variable t does not occur explicitly in the right-hand side of (5–1–1) the equations are said to be *autonomous* or *stationary*; otherwise they are non-autonomous or non-stationary.

If the functions f_i in (5–1–1) satisfy certain conditions then there exists a unique solution to (5–1–1) which passes through a given point $x^0 = (x_1^0, \ldots, x_n^0)'$ at a given time t_0. One such condition is that the f_i should be continuous and have continuous first-partial derivatives $\frac{\partial f_i}{\partial x_j}$, although this is by no means the weakest possible condition [see, e.g., Lefschetz (1965)]. Thus for each value of t a solution gives values for each x_i and describes a curve in the space formed by the x_i and t; this is called an *integral curve*. The solution can also be regarded as a curve in the space of the x_i only and is then known as a *trajectory*. Thus trajectories are the projections of integral curves on to the space of the x_i. For autonomous systems the trajectories are independent of a translation in time and are often called *paths*.

Example 5–1–1
Consider the equations

$$\dot{x}_1 = x_2$$
$$\dot{x}_2 = -x_1 + 2x_2$$

with $x_1(0) = 0$, $x_2(0) = 1$. The solution is easily found to be

$$x_1 = te^t$$
$$x_2 = (t + 1)e^t \tag{5–1–5}$$

so that the integral curve in this case looks as in Fig. 5–1. On eliminating t from

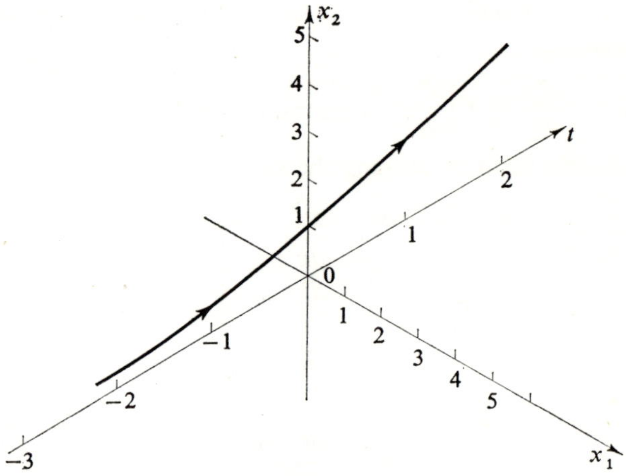

Fig. 5–1

(5–1–5) $\left(\text{or directly from the fact that } \dfrac{dx_1}{dx_2} = \dfrac{x_2}{-x_1 + 2x_2}\right)$ it is seen that $x_1 = (x_2 - x_1) \ln (x_2 - x_1)$ so that the path through $x_1 = 0$, $x_2 = 1$ is as shown in Fig. 5–2.

Values of x_i, $i = 1, 2, \ldots, n$, which make the right-hand side of (5–1–1) vanish for all $t \geqslant t_0$ are known as *critical* or *equilibrium* points, and a trajectory starting at an equilibrium point will remain there for all subsequent time.

Generally speaking stability is concerned with whether a system will remain in an equilibrium state when slightly disturbed from it or if it will move further away from that state. More precise mathematical definitions of stability due to Liapunov are now given.

The solution of (5–1–1) which passes through the point x^0 at time t_0 will be written $x(t, x^0, t_0)$ and it will be assumed that the origin is a critical point of (5–1–1), i.e., $f(0, t) = 0$, $t \geqslant t_0$. (A critical point at $x = a$ can always be transferred to the origin by means of the translation $x^* = x - a$.)

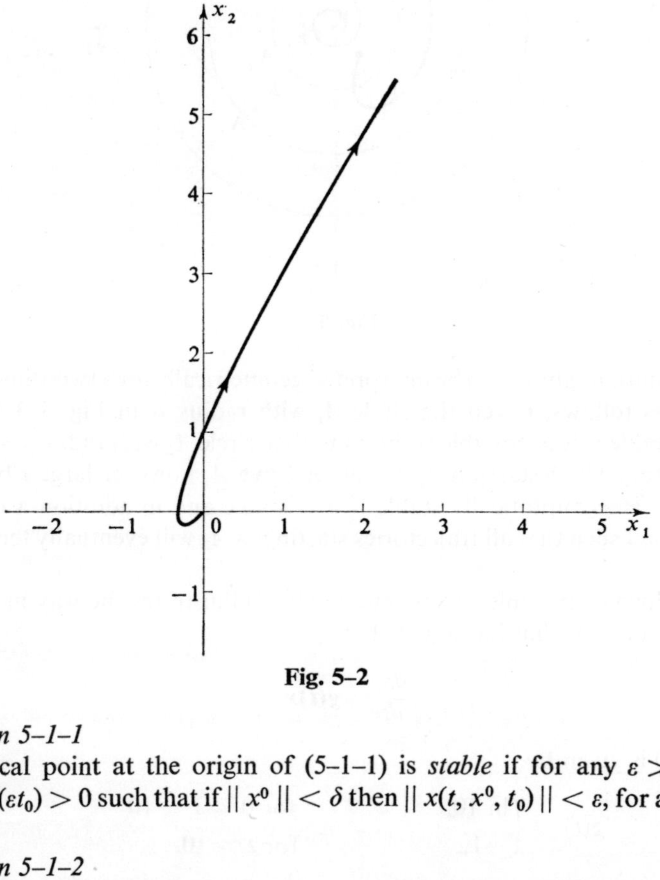

Fig. 5–2

Definition 5–1–1
The critical point at the origin of (5–1–1) is *stable* if for any $\varepsilon > 0$ there exists a $\delta(\varepsilon t_0) > 0$ such that if $\| x^0 \| < \delta$ then $\| x(t, x^0, t_0) \| < \varepsilon$, for all $t \geqslant t_0$.

Definition 5–1–2
The critical point at the origin of (5–1–1) is *quasi-asymptotically stable* if there exists a $\delta^* > 0$ such that $\| x^0 \| < \delta^*$ implies $\lim_{t \to \infty} \| x(t, x^0, t_0) \| = 0$.

Definition 5–1–3
If the origin of (5–1–1) is both stable and quasi-asymptotically stable then it is said to be *asymptotically stable.*

For asymptotic stability it is not enough that all trajectories tend to the origin with increasing time – the origin must also be stable.

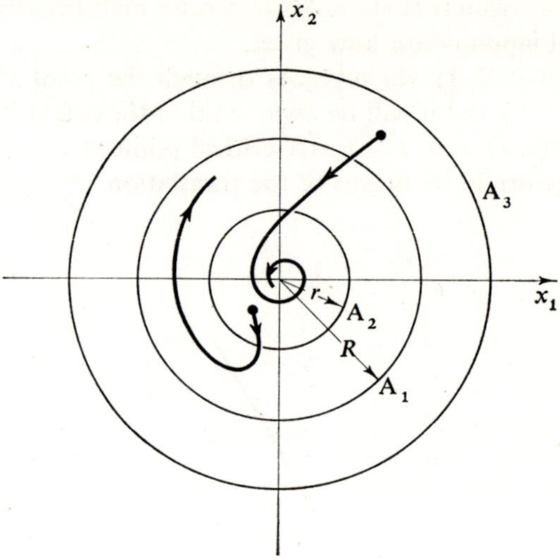

Fig. 5–3

The definitions above can be interpreted geometrically for a two-dimensional problem as follows. Given the circle A_1 with radius R in Fig. 5–3 then the origin is stable if it is possible to find another circle A_2 with radius r such that any trajectory which starts in A_2 will never leave A_1 however large t becomes. The origin is asymptotically stable if it is stable and in addition a circle A_3 can be found such that all trajectories starting in A_3 will eventually tend to the origin.

The following example due to Zubov (1962) illustrates the way in which δ depends on ε in Definition 5–1–1. Let

$$\frac{dx}{dt} = g(t)x \tag{5–1–6}$$

where $g(t)$ is given by

$$g(t) = \begin{cases} \ln 10, & \text{for } 0 \leqslant t \leqslant 10 \\ -1, & \text{for } t > 10. \end{cases}$$

Suppose $x = x^0$ when $t = 0$, then the solution of (5–1–6) is

$$x = \begin{cases} 10^t x^0, & \text{for } 0 \leqslant t \leqslant 10 \\ 10^{10} e^{(10-t)} x^0, & \text{for } t > 10. \end{cases} \tag{5–1–7}$$

This is evidently asymptotically stable. If, however, the numerical value of x^0 changes by only 10^{-5} the solution (5–1–7) changes by $10^5 e^{-10} \approx 3$ at $t = 20$.

This example illustrates the theoretical nature of Liapunov's definitions of stability. Many systems that are unstable according to this definition perform quite adequately from a practical point of view. Conversely there are systems which although stable according to the definitions would not behave well in practice. These considerations have led to an alternative definition of stability known as 'practical stability' in which not only the allowable excursions in the system response but also the allowable range of initial conditions are specified in advance [see, e.g., La Salle and Lefschetz (1961)].

The constant δ in Definition 5–1–1 also depends on t_0 in general and in the case when δ depends on ε only and not on the initial instant the stability is said to be *uniform* [see, e.g., Hahn (1963)].

Systems which are not stable are generally said to be *unstable* but the following definition is more specific.

Definition 5–1–4

The critical point at the origin of (5–1–1) is said to be *unstable* if for some $\varepsilon > 0$ there exists a set of numbers τ_k, $k = 1, 2, \ldots$ which tend to ∞ as $k \to \infty$ and a sequence of initial points x_k, $k = 1, 2, \ldots$ which tend to 0 as $k \to \infty$, such that $\| x(t_0 + \tau_k, x_k, t^0) \| \geqslant \varepsilon$ for all k.

The origin is said to be *completely unstable* if for some $\varepsilon > 0$ the distance from the origin of every trajectory $x(t, x^0, t_1)$ becomes greater than ε after some finite time. It is supposed that $t_1 \geqslant t_0$ and $0 < \| x^0 \| < \varepsilon$.

The above definitions of stability will be sufficient for the purposes of this book and for other definitions the reader is referred, for example, to Hahn (1963).

5–2 Liapunov functions

To investigate the stability of a system of differential equations without having to solve them Liapunov devised his so-called 'second method' in 1892 [Liapunov (1907)]. The idea involved is a generalization of the concept of energy for a conservative dynamic system. In such a system the energy is a positive function which decreases to zero as an equilibrium state is approached.

If then, for a general system, a function with properties similar to those of an energy function can be found the stability of the system can be guaranteed. That is to say, if a function $V(x)$ can be constructed such that within some region around the critical point the contours of constant $V(x)$ represent concentric shells which decrease to zero as $\| x \| \to 0$ and if the trajectories of the system cross these shells in an inward direction then the critical point is asymptotically stable.

Functions with the above properties are known as *Liapunov functions* and we proceed now with formal definitions of these functions and with statements of the Liapunov stability theorems.

Definition 5–2–1

A function $V(x)$ is said to be *positive* (*negative*) *definite* if $V(0) = 0$ and in some neighbourhood $\| x \| \leqslant h$ of the origin $V(x) > 0$ (< 0) for $x \neq 0$.

If the strict inequality signs above are replaced by $\geqslant$ ($\leqslant$) then $V(x)$ is said to be positive (negative) semi-definite.

Definition 5–2–2
A function $V(x, t)$ defined for $\| x \| \leqslant h$, $t \geqslant t_0$ is positive (negative) definite if $V(0, t) = 0$ and if, for $\| x \| \leqslant h_1 \leqslant h$, a positive definite function $W(x)$ exists such that

$$V(x, t) \geqslant W(x) \qquad [V(x, t) \leqslant -W(x)].$$

$V(x, t)$ is positive (negative) semi-definite if $V(0, t) = 0$ and if $V(x, t) \geqslant 0$ ($\leqslant 0$) for $\| x \| \leqslant h_1 \leqslant h$ and $t \geqslant t_0$.

Definition 5–2–3
A function $V(x, t)$ is *decrescent* if there exists a positive definite function $U(x)$ such that, for $\| x \| < h$, $t \geqslant t_0$,

$$\| V(x, t) \| \leqslant U(x).$$

Exercise 5–2–1
An alternative definition of a decrescent function is as follows. The function $V(x, t)$ is decrescent if

$$\lim_{\|x\| \to 0} V(x, t) = 0$$

holds uniformly in t. Show the equivalence of these two definitions.

Definition 5–2–4
A function $V(x, t)$ which is defined for $\| x \| \leqslant h$, $t \geqslant t_0$, and which is positive definite is said to be a Liapunov function for system (5–1–1) provided that $\dot{V}$ is negative semi-definite. Here $\dot{V}$ is the total derivative of V with respect to (5–1–1) and is given by

$$\dot{V} = \frac{dV}{dt} = \frac{\partial V}{\partial t} + \sum_{i=1}^{n} \frac{\partial V}{\partial x_i} f_i. \qquad (5\text{–}2\text{–}1)$$

[Sometimes rather wider definitions of Liapunov functions are given, e.g., Hahn (1963), p. 15.]

Theorem 5–2–1
The origin of (5–1–1) is stable if there exists a Liapunov function $V(x, t)$ in some region R: $\{\| x \| \leqslant h,\ t \geqslant t_0\}$.

Theorem 5–2–2
The origin of (5–1–1) is asymptotically stable if there exists a positive definite decrescent function $V(x, t)$ which has a negative definite total derivative (5–2–1) with respect to (5–1–1).

Theorem 5–2–3 (*Instability*)
The origin of (5–1–1) is unstable if there exists a decrescent function $V(x, t)$ with a negative definite total derivative $\dot{V}$ with respect to (5–1–1) and, for all sufficiently large t, there are points in every neighbourhood of the origin at which $V < 0$.

Theorem 5–2–4 (*Autonomous systems*)
If system (5–1–1) is autonomous then its origin is stable provided there exists a Liapunov function $V(x)$ [i.e., $V(x)$ positive definite, $\dot{V}$ negative semi-definite]. The origin is asymptotically stable if there exists a positive definite $V(x)$ which has negative definite total derivative $\dot{V}$ with respect to (5–1–1).

Theorem 5–2–5 (*Barbashin*)
If system (5–1–1) is autonomous the condition that $\dot{V}$ should be negative definite can be replaced by the weaker conditions:

(*a*) $\dot{V}$ is negative semi-definite
(*b*) $\dot{V}$ does not vanish identically on any non-trivial trajectory of (5–1–1).

Theorem 5–2–6
The origin of (5–1–1) is *asymptotically stable in the whole* (i.e., for trajectories starting from all points of phase space) if a function $V(x, t)$ exists which has the following properties throughout the whole phase space:

(*a*) $V(x, t)$ is positive definite;
(*b*) $V(x, t)$ is decrescent;
(*c*) Given any $\varepsilon > 0$ there exists a $\delta > 0$ such that $V(x, t) > \varepsilon$ whenever $\| x \| > \delta$ and $t \geqslant t_0$ [such a $V(x, t)$ is said to be ***radially unbounded***];
(*d*) $\dot{V}$ with respect to (5–1–1) is negative definite.

Theorem 5–2–7
Trajectories of (5–1–1) which start from a region D containing the origin will be asymptotically stable if there exists a function $V(x, t)$ with the following properties:

(*a*) $V(x, t)$ is positive definite in D;
(*b*) $\dot{V}$ with respect to (5–1–1) is at least negative semi-definite in D;
(*c*) $\dot{V}(x, t) \not\equiv 0$ on any trajectory of the system in D other than $x = 0$;
(*d*) $\nabla V(x, t) \not\equiv 0$ in D except for $x = 0$;
(*e*) one of the level surfaces $V(x, t) =$ constant bounds the region D.

The above theorems are still true if positive and negative are interchanged.

The region D is called a region of *asymptotic stability* and if it forms the entire set of initial conditions giving rise to asymptotically stable trajectories it is called the *domain of attraction*. When D is the whole phase space asymptotic stability in the whole results. A distinction is sometimes made between asymptotic stability in the whole and *asymptotic stability in the large*; the latter being reserved for cases where Eqn (5–1–1) is not defined for the whole of the phase space. Asymptotic stability in the whole is often called *complete stability*.

It is natural to inquire under what conditions the converses of the various theorems given above hold. Investigation of the converse theorems is more difficult than that of the theorems themselves and a full theory is not yet available. For completeness, however, we give one theorem on the existence of a Liapunov function.

Theorem 5–2–8

Let the components f_i of the right side of (5–1–1) and their partial derivatives $\frac{\partial f_i}{\partial x_j}$, $i, j, = 1, 2, \ldots, n$, be continuous in some region R and let the origin of (5–1–1) be stable. Then a positive definite function $V(x, t)$ exists in R and has continuous partial derivatives with respect to all of its variables and a negative semi-definite total derivative (5–2–1) with respect to (5–1–1).

For proofs of the above theorems and extensions to them we refer the reader to Hahn (1963), La Salle and Lefschetz (1961), Zubov (1964), and Lehnigk (1966).

The following simple example will illustrate the use of Liapunov functions for an autonomous system.

Example 5–2–1

Consider the system

$$\frac{dx_1}{dt} = x_1 - x_1x_2$$
$$\frac{dx_2}{dt} = x_1x_2 - x_2 \qquad (5\text{–}2\text{–}2)$$

which has critical points at the origin and at (1, 1). For the critical point at the origin set

$$2V = x_1^2 - x_2^2$$

so that

$$\dot{V} = x_1(x_1 - x_1x_2) - x_2(x_1x_2 - x_2)$$
$$= x_1^2 + x_2^2 - x_1^2x_2 - x_1x_2^2 \qquad (5\text{–}2\text{–}3)$$

From (5–2–3) $\dot{V}$ is positive definite for sufficiently small x_1,x_2 but V itself can be positive for points arbitrarily near the origin. Thus by Theorem 5–2–3 the origin is unstable.

Now write

$$\xi_1 = x_1 - 1, \qquad \xi_2 = x_2 - 1$$

so that (5–2–2) becomes

$$\begin{aligned}\frac{d\xi_1}{dt} &= -\xi_1\xi_2 - \xi_2 \\ \frac{d\xi_2}{dt} &= \xi_1\xi_2 + \xi_1\end{aligned} \tag{5–2–4}$$

which again has a critical point at the origin. The function

$$\begin{aligned}V &= \xi_1 + \xi_2 - \ln(1 + \xi_1) - \ln(1 + \xi_2) \\ &= \tfrac{1}{2}\xi_1^2 + \tfrac{1}{2}\xi_2^2 + \cdots\end{aligned} \tag{5–2–5}$$

is positive definite sufficiently near the origin. The total derivative of V with respect to (5–2–4) is given by

$$\begin{aligned}\dot{V} &= \dot{\xi}_1 + \dot{\xi}_2 - \frac{\dot{\xi}_1}{1 + \xi_1} - \frac{\dot{\xi}_2}{1 + \xi_2} \\ &= -\xi_1\xi_2 - \xi_2 + \xi_1\xi_2 + \xi_1 + \xi_2 - \xi_1 \\ &= 0.\end{aligned}$$

Thus V is a Liapunov function for (5–2–4) and hence the critical point at (1, 1) of (5–2–2) is stable.

Exercise 5–2–2
Show that V given by (5–2–5) is a first integral of (5–2–4). Plot roughly the phase portrait of system (5–2–2).

Exercise 5–2–3
Investigate the use of the function $V = \xi_1^2 + \xi_2^2$ for system (5–2–4).

In addition to their use for stability analysis Liapunov functions can be used to measure the rate at which system trajectories actually approach the origin. In this connection the following theorem [Kalman and Bertram (1960)] is useful.

Theorem 5–2–9
If the origin of the system (5–1–1) is uniformly asymptotically stable in the large then if $V(x, t)$ is a Liapunov function for (5–1–1),

$$V[x(t, x^0, t_0)] \leqslant \exp\{-k(t - t_0)\}\, V(x^0, t_0)$$

where k is the minimum value of the ratio $-\dot{V}(x, t)/V(x, t)$.

Generally speaking the reciprocal of k can be regarded as the largest time constant of the system in some region of state space. Care must be taken in using the above estimate of the distance of the system from the origin in cases when $\dot{V}(x, t)$ is only semi-definite.

5–3 Linear systems

We now consider linear autonomous systems in which case Eqn (5–1–1) reduces to

$$\dot{x} = Ax \tag{5–3–1}$$

where A is a real constant $n \times n$ matrix. When Eqn (5–1–3) is linear, i.e., of the form

$$\frac{d^n x}{dt^n} + c_1 \frac{d^{n-1}x}{dt^{n-1}} + \cdots + c_{n-1}\frac{dx}{dt} + c_n x = 0 \tag{5–3–2}$$

then using the substitutions of (5–1–4), (5–3–2) can be written in the form (5–3–1) with A in companion form (3–2–1). Suppose that Eqn (5–3–1) satisfies the initial condition $x(t_0) = x^0$; then it is easily verified (from the properties of the matrix exponential, Chapter 4) that its general solution is

$$x(t, x^0, t_0) = e^{A(t-t_0)}x^0.$$

The following theorems are fundamental in the study of linear autonomous systems [for proofs see, e.g., Hahn (1963) or Willems (1970)].

Theorem 5–3–1

For Eqn (5–3–1) the origin is asymptotically stable if and only if all the characteristic roots of A have negative real parts.

This result leads to the classical works of Maxwell, Hermite, Routh, and Hurwitz which give necessary and sufficient conditions, in terms of the elements of A (especially for the case where A is in companion form), for the real parts of the characteristic roots of A all to be negative. For details of these conditions and their relationship with the later work of Bilharz the reader is referred to the papers by Parks (1963, 1969) and Anderson (1967), and books by Lehnigk (1966) and Marden (1966).

The second method of Liapunov takes a particularly simple form when applied to the system (5–3–1) and gives rise to the following important theorem.

Theorem 5–3–2

For Eqn (5–3–1) the origin is asymptotically stable if and only if there exists a symmetric positive definite matrix P which is the unique solution of the matrix equation

$$A'P + PA = -Q \tag{5–3–3}$$

for any given symmetric positive definite matrix Q. Notice that $V = x'Px$ is a Liapunov function for (5–3–1) with

$$\begin{aligned}\dot{V} &= x'(A'P + PA)x \\ &= -x'Qx.\end{aligned} \tag{5–3–4}$$

Also important is the following theorem on the instability of (5–3–1).

Theorem 5–3–3
The null solution of (5–3–1) is unstable if at least one of the characteristic roots of A has positive real part.

If all the characteristic roots of A have negative real parts then A is called a *stability* matrix. We will investigate some aspects of the structure of stability matrices in Chapter 7. If either A is a stability matrix or at least one of its characteristic roots has positive real part then (5–3–1) is said to have *significant stability behaviour*. If Eqn (5–3–1) does not have significant stability behaviour it is said to have *critical stability behaviour* and the following apply:

Theorem 5–3–4
The origin is stable for Eqn (5–3–1) if and only if the real parts of the characteristic roots of A are non-positive and the elementary divisors corresponding to those characteristic roots of A with zero real part are linear.

Theorem 5–3–5
The origin is unstable for Eqn (5–3–1) if A has a characteristic root with zero real part to which there corresponds a non-linear elementary divisor.

For further results on the critical cases the reader is referred to Liapunov (1966), Lehnigk (1966), Hahn (1963), and Zubov (1964).

Theorem 5–3–2 is particularly important for our purposes in this book and the solution of Eqn (5–3–3) is taken up in detail in the next chapter. This equation is known as a *Liapunov matrix equation*. With regard to the use of the Liapunov matrix equation in stability analysis (it has other uses as will be seen later) the following remarks are important.

If Q is positive semi-definite rather than positive definite and P is positive definite the origin is stable by Theorem 5–2–4. It may or may not, however, be asymptotically stable [Lehnigk (1967)]. Thus

$$A = \begin{bmatrix} -1 & 0 \\ 0 & -2 \end{bmatrix}, \quad P = \begin{bmatrix} \frac{3}{2} & 2 \\ 2 & 3 \end{bmatrix}, \quad Q = 3\begin{bmatrix} 1 & 2 \\ 2 & 4 \end{bmatrix}$$

exhibits asymptotic stability with Q positive semi-definite and P positive definite, whereas

$$A = \begin{bmatrix} 0 & 1 \\ 0 & -1 \end{bmatrix}, \quad P = \begin{bmatrix} 1 & 1 \\ 1 & 2 \end{bmatrix}, \quad Q = \begin{bmatrix} 0 & 0 \\ 0 & 2 \end{bmatrix}$$

does not exhibit asymptotic stability although P is positive definite and Q is positive semi-definite (A has characteristic roots 0 and 1).

The example

$$A = \begin{bmatrix} 1 & -3 \\ 2 & -4 \end{bmatrix}, \quad P = \begin{bmatrix} 1 & -1 \\ -1 & 1 \end{bmatrix}, \quad Q = \begin{bmatrix} 2 & -2 \\ -2 & 2 \end{bmatrix}$$

shows that even though A is stable (its roots are -1 and -2) and Q is positive semi-definite, P is only positive semi-definite. In this case, of course, the quadratic form $x'Px$ is not a Liapunov function for $\dot{x} = Ax$.

To conclude this section we give a useful theorem due to Lehnigk (1966).

Theorem 5–3–6

Let the origin of the linear system (5–3–1) be asymptotically stable and assume $A \neq kI_n$ (with k a real constant). Then if B is a singular matrix such that $B(\exp At)x^0$ is not identically zero for all $x^0 \neq 0$ there exists a positive definite solution P of the matrix equation

$$A'P + PA = -B'B.$$

In this section the discussion has been restricted for simplicity to linear systems with real coefficients. For extension to the case where the elements of A are complex the reader is referred to Lehnigk (1966).

Exercise 5–3–1

Solve the system of linear differential equations

$$\frac{dx_1}{dt} = 2x_1 + x_2 - 6x_3 - 6x_4$$

$$\frac{dx_2}{dt} = x_2$$

$$\frac{dx_3}{dt} = -3x_1 - x_2 + 5x_3 + 6x_4$$

$$\frac{dx_4}{dt} = 3x_1 + x_2 - 6x_3 - 7x_4$$

by reducing the matrix of coefficients to its Jordan form.

Exercise 5–3–2

Verify that the solution of the matrix differential equation

$$\frac{dX}{dt} = AX + XB, \quad X(0) = C$$

is $X = e^{At}Ce^{Bt}$. Use this result to show that the solution of

$$AX + XB = C$$

is given by

$$X = -\int_0^\infty e^{At}Ce^{Bt}\,dt.$$

5–4 Discrete linear systems

The discrete analogue of Eqn (5–3–1) is the difference equation

$$x(t_{k+1}) = Ax(t_k),\ k = 0, 1, 2, \ldots \tag{5–4–1}$$

and many of the definitions of Section 5–1 carry over to such difference equations with only slight changes [see Hahn (1963), p. 146; Willems (1970)].

In particular it is possible to use the quadratic form

$$V = x'Px \tag{5–4–2}$$

as a Liapunov function in the following way. Instead of $\dot{V}$ we form

$$\Delta V = V(t_{k+1}) - V(t_k), \tag{5–4–3}$$

the *difference* of V with respect to (5–4–1) to give

$$\Delta V = x'(A'PA - A)x \tag{5–4–4}$$

$$= -x'Qx \tag{5–4–5}$$

This leads to an analogue of Theorem 5–3–2.

Theorem 5–4–1

The null solution (the origin) of (5–4–1) is asymptotically stable if and only if there exists a symmetric positive definite matrix P which is the unique solution of the matrix equation

$$A'PA - P = -Q \tag{5–4–6}$$

for any given symmetric positive definite matrix Q. [Eqn (5–4–6) is the Liapunov matrix equation for the discrete system (5–4–1).]

Also notice that the general solution of (5–4–1) with initial condition

$$x(t_{k_0}) = x^0$$

is

$$x(t_k, x^0, t_{k_0}) = A^{k-k_0} x^0$$

and that Theorems 5–3–1, 5–3–3, 5–3–4, and 5–3–5 hold for system (5–4–1) provided that the word 'characteristic root' is everywhere replaced by 'logarithm of characteristic root' in their statements.

As might be expected there is a close connection between the solution of (5–3–3) and that of (5–4–6). This is easily seen by putting

$$A' = (B + I)^{-1}(B - I) \tag{5–4–7}$$

or

$$B = (I + A')(I - A')^{-1} \tag{5–4–8}$$

into (5–3–3) which then becomes

$$2BPB' - 2P = -(B + I)Q(B' + I) \tag{5–4–9}$$

or

$$B'PB - P = -2(I - A')^{-1}Q(I - A)^{-1}. \tag{5–4–10}$$

This transformation from (5–3–3) to (5–4–6) can also be carried out when A is complex [for details see Smith (1966)].

The transformation represented by (5–4–7) and (5–4–8) also assumes importance in the solution of the Liapunov matrix equation in terms of an infinite series. This will be discussed in detail in the next chapter.

5–5 The construction of Liapunov functions

It has already been seen that for linear systems a quadratic form can always be tried as a Liapunov function and, indeed, much of this book is concerned with this particular application. Another special case in which a Liapunov function is available is the so-called 'Luré problem.' Here linear systems with certain added non-linear terms are considered – the system equations being of the form

$$\begin{aligned} \dot{y} &= Ay + b\xi \\ \dot{\xi} &= f(\sigma) \\ \sigma &= c'y - \rho\xi \end{aligned} \tag{5–5–1}$$

where A is an $n \times n$ stable matrix, y and b are column vectors, c' is a row vector, and ξ, σ, ρ are scalars. The non-linear term $f(\sigma)$ is assumed to satisfy the following conditions:

(*a*) $f(\sigma)$ is continuous for all σ;
(*b*) $\sigma f(\sigma) > 0$ for all $\sigma \neq 0$;
(*c*) $f(0) = 0$. (5–5–2)

On writing

$$\begin{aligned} x &= Ay + b\xi \\ \sigma &= c'y - \rho\xi \end{aligned} \tag{5–5–3}$$

(5–5–1) is transformed into

$$\begin{aligned} \dot{x} &= Ax + bf(\sigma) \\ \dot{\sigma} &= c'x - \rho f(\sigma) \end{aligned} \tag{5–5–4}$$

provided that $\rho + c'A^{-1}b \neq 0$; this latter condition also ensures that (5–5–1) has a critical point at the origin. Now consider

$$V = x'Bx + \int_0^\sigma f(\sigma)d\sigma \tag{5–5–5}$$

as a possible Liapunov function. An easy calculation shows that the total derivative of V with respect to (5–5–4) is given by

$$\dot{V} = x'(A'B + BA)x - \rho f^2(\sigma) + 2f(\sigma)\{\tfrac{1}{2}(b'Bx + x'Bb) + \tfrac{1}{2}c'x\} \tag{5–5–6}$$

It now follows from the Sylvester conditions that $\dot{V}$ will be negative definite provided

$$\rho > (Bb + \tfrac{1}{2}c)' C^{-1} (Bb + \tfrac{1}{2}c) \tag{5–5–7}$$

where B is the positive definite solution of

$$A'B + BA = -C \tag{5–5–8}$$

for some positive definite C (A is stable). Thus if the inequality (5–5–7) holds the system (5–5–1) is completely stable. For further details of the Luré problem and its ramifications the reader is referred to Lefschetz (1965), and Aizerman and Gantmacher (1964).

Apart from the two cases quoted above there is, unfortunately, no certain method of finding a Liapunov function for a general non-linear problem. A number of suggestions have been made, however, for such construction in the general case and we conclude this chapter with a very brief introduction to some of these techniques.

5–5–1 Zubov's method [Zubov (1962), Vogt (1964)]

A Liapunov function can be found for the autonomous system $\dot{x} = f(x)$ by solving the partial differential equation

$$\frac{dV}{dt} = f' \text{ grad } V(x) = -\phi(x) \tag{5–5–9}$$

where $\phi(x)$ is some suitable positive definite function. The following results apply (when the first approximation to $\dot{x} = f(x)$ is asymptotically stable):

(*a*) Each $\phi(x)$ determines a unique set of closed non-intersecting hypersurfaces $V(x) = c$ about the origin, where c is constant and $0 < c < \infty$. For different $\phi(x)$ these hypersurfaces coincide only at $c = 0$ and $c = \infty$ and if $0 < c_1 < c_2 < \infty$ then $V(x) = c_1$ lies entirely within $V(x) = c_2$.

(*b*) The domain of asymptotic stability D is given by $0 \leqslant V(x) < \infty$.

(*c*) If the limit of a sequence $\{x_n\}$ of points in D tends to a point on the boundary of D then $V(x_n) \to \infty$ as $n \to \infty$.

(*d*) A first approximation to $V(x)$ is the quadratic form $x'Bx$ where B is the (real, symmetric, positive definite) solution of

$$A'B + BA = -C$$

where $\dot{x} = Ax$ is the first approximation to $\dot{x} = f(x)$ and $x'Cx$ the first approximation to $\phi(x)$.

(*e*) If $x^0 \in D$ then

$$V(x) = \int_0^\infty \phi\{x(t, x^0, 0)\} dt.$$

(*f*) If $x^0 \in D$ then

$$V\{x(t, x^0, 0)\} = V(x^0) - \int_0^t \phi\{x(t, x^0, 0)\}dt.$$

(*g*) If the substitution $V^* = 1 - \exp[-V(x)]$ is made an alternative form of the method is obtained in which the partial differential equation to be solved is

$$\dot{V}^* = f' \text{ grad } V^* = -\phi(x)\{1 - V^*(x)\}$$

and the boundary of the domain of asymptotic stability is then given by $V^*(x) = 1$.

When $V(x)$ or [or $V^*(x)$] cannot be obtained in closed form the region of asymptotic stability cannot be obtained exactly but a convergent power series can be obtained from which estimates of D are found. Unfortunately the convergence to the exact stability region is not uniform – a fact which makes the method less attractive than it might otherwise have been.

An interesting generalization of Zubov's method can be found in Szegö (1962). [See also Szegö and Geiss (1963).]

5–5–2 Ingwerson's method [Ingwerson (1961)]

On differentiation the autonomous system $\dot{x} = f(x)$ becomes $\ddot{x} = A(x)\dot{x}$ where $A(x)$ is the jacobian matrix $\left(\frac{\partial f_i}{\partial x_j}\right)$. When A is constant then $V = x'Bx$ is a Liapunov function for the system provided the solution B of $A'B + BA = -C$ is positive definite for some symmetric positive definite C. In this case the elements of B are given by $\frac{1}{2}\left(\frac{\partial^2 V}{\partial x_i \partial x_j}\right)$ and V could be obtained by carrying out two integrations. Since the matrix equation can equally be solved for non-constant matrices as for constant ones the following formal procedure was suggested by Ingwerson:

(*a*) Derive the jacobian $A(x)$. With this $A(x)$ obtain the solution $B(x)$ of the matrix equation for some positive definite C.

(*b*) Form a new matrix $B(x_i, x_j)$ from B by setting to zero all terms other than x_i and x_j in the element $b_{ij}(x)$.

(*c*) Carry out the integration

$$\nabla V = \int_0^x B(x_i, x_j)\, dx$$

over each component of x treating the other components as constant. This results in the vector ∇V.

(*d*) Evaluate the line integral

$$V = \int_0^x \nabla V \, . \, dx$$

along any path of integration. Because of the way in which $B(x_i, x_j)$ has been chosen the line integral above is independent of the path of integration and the simplest possible path may be used. Thus we can take

$$V(x) = \int_0^{x_1} \nabla V_1(y_1, 0, \ldots, 0) dy_1 + \int_0^{x_2} \nabla V_2(x_1, y_2, \ldots, 0)\, dy_2 + \cdots + \int_0^{x_n} \nabla V_n(x_1, x_2, \ldots, x_{n-1}, y_n)\, dy_n$$

where ∇V_i is the ith component of ∇V.

(*e*) Test V to see if it is a Liapunov function.

The following example due to Ingwerson (1961) will make the details of the method clear.

Example 5–5–1

$$\dot{x}_1 = x_2$$

$$\dot{x}_2 = x_3$$

$$\dot{x}_3 = -(x_1 + cx_2)^3 - bx_3$$

$$A(x) = \begin{bmatrix} 0 & 1 & 0 \\ 0 & 0 & 1 \\ -3(x_1 + cx_2)^2 & -3c(x_1 + cx_2)^2 & b \end{bmatrix}$$

$$B(x) = \begin{bmatrix} 3b(x_1 + cx_2)^2 & 3(x_1 + cx_2)^2 & 0 \\ 3(x_1 + cx_2)^2 & b^2 + 3c(x_1 + cx_2)^2 & b \\ 0 & b & 1 \end{bmatrix}.$$

This gives

$$B(x_i, x_j) = \begin{bmatrix} 3bx_1^2 & 3(x_1 + cx_2)^2 & 0 \\ 3(x_1 + cx_2)^2 & b^2 + 3c^3x_2^2 & b \\ 0 & b & 1 \end{bmatrix}$$

$$\nabla V = \begin{bmatrix} bx_1^3 + \frac{1}{c}(x_1 + cx_2)^3 - \frac{1}{c}x_1^3 \\ (x_1 + cx_2)^3 + b^2x_2 + bx_3 \\ bx_2 + x_3 \end{bmatrix}$$

and finally,

$$V = \int_0^x \nabla V \, . \, dx$$

$$= \frac{b}{4} x_1 + \frac{1}{4c}(x_1 + cx_2)^4 - \frac{1}{4c} x_1^4 + \frac{b^2}{2} x_2^2 + bx_2x_3 + \frac{1}{2} x_3^2.$$

It follows that if b, c, and $bc - 1$ are positive V is positive definite, and under the same conditions

$$\dot{V} = \nabla V \, . \, \dot{x} = -(bc - 1)(3x_1^2 + 3cx_1x_2 + c^2x_2^2)x_2^2$$

is negative semi-definite. The origin of the system is therefore stable.

5–5–3 The variable gradient method [Schultz and Gibson (1962)]

This method which is a generalization of that of Ingwerson proceeds as follows. Let

$$\dot{x} = A(x)x, \qquad x(0) = 0$$

where

$$A(x) = \begin{bmatrix} 0 & 1 & 0 & \cdots & 0 \\ 0 & 0 & 1 & \cdots & 0 \\ \cdot & \cdot & \cdot & \cdots & \cdot \\ \cdot & \cdot & \cdot & \cdots & \cdot \\ \cdot & \cdot & \cdot & \cdots & \cdot \\ a_1(x) & a_2(x) & a_3(x) & \cdots & a_n(x) \end{bmatrix}$$

i.e., the system is supposed to be in companion form but the coefficients $a_i(x)$ are functions of x.

Start with an arbitrary column vector

$$\nabla V = Bx$$

and impose the following restrictions on the matrix B

(a) $b_{nn} = 2$;
(b) $b_{ii} > 0$ and functions of x_i only;
(c) $b_{ij} =$ constant $+$ function of $x_i, \ldots, x_{n-1}$; $i \neq j$.

These restrictions are meant to increase the possibility of V being positive definite. Then form $\dot{V} = \nabla V' \dot{x}$ and make it negative definite (or semi-definite) by suitably choosing the elements b_{ij}. Now if curl (∇V) is set equal to zero a set of $\frac{1}{2}n(n - 1)$ equations is obtained which allow the elements b_{ij} to be fully determined (not, however, uniquely). The vanishing of curl ∇V

is just the condition that ∇V should be the gradient of some scalar function V. [Weatherburn (1951).]

A straightforward line integration

$$V = \int_0^x \nabla V . dx$$

will enable the scalar function V to be determined as in Ingwerson's method. Finally, a check must be carried out to ensure that $\dot{V}$ has not been altered by using the curl relations.

5–5–4 Krasovskii's method [Krasovskii (1957)]

This method relies on the following result. The function

$$V(x) = f'Bf \qquad [f' = (f_1, f_2, \ldots, f_n)]$$

is a Liapunov function for $\dot{x} = f(x)$, $f(0) = 0$ provided that the symmetric matrix

$$C(x) = -\{A'(x)B + BA(x)\}$$

is positive definite for all x where $A(x) = \left[\frac{\partial f_i}{\partial x_j}\right]$ and B is an arbitrary positive definite matrix.

It can be shown that if, as before, $\dot{x} = Ax$ is the first approximation to the system $\dot{x} = f(x)$ and the former is asymptotically stable then a Liapunov function for the latter is $f'Bf$ where B is the solution of the constant matrix equation $A'B + BA = -C$ for some positive definite C.

These are by no means the only available methods for constructing Liapunov functions. For example, Infante and Clarke (1965) describe an interesting technique in which they set up a new system which has an energy integral and which is 'close' in some sense to the system under investigation, the hope being that the energy integral (with some modifications, perhaps) can be used as a Liapunov function for the original system. For a comprehensive review of methods of constructing Liapunov functions the reader is referred to Gurel and Lapidus (1968).

Many, but not all, of the methods available for the construction of Liapunov functions can be converted to digital computing techniques. Particularly suited to computing algorithms are the methods of Krasovskii, Ingwerson, and Zubov. A critical comparison of these techniques for second-order systems, together with other numerical non-Liapunov methods can be found in Hewit and Storey (1969) [see also Hewit (1968)]. Hewit also discusses the method of Rodden (1964) for plotting the boundary of the area of asymptotic stability for a given Liapunov function. Finally, we mention that the arbitrary parameters available in the construction procedures of Krasovskii, Ingwerson, and Zubov (i.e., the matrix C and the function ϕ) can be adjusted by optimization programmes to obtain the maximum area of asymptotic stability [Hewit and Storey (1967, 1969)].

Discussion and further reading

This chapter has dealt briefly with the fundamental definitions and theorems of Liapunov's stability theory. A short account has also been given of the stability theory of continuous and discrete linear systems. For general works on the theory of ordinary differential equations the reader is referred to Coddington and Levinson (1955), Hartman (1964), and Lefschetz (1962). For further details of Liapunov theory Zubov (1962, 1964), La Salle and Lefschetz (1961), Lehnigk (1966), Krasovskii (1963), and Hahn (1963) should be consulted. The reviews by Brockett (1966) and Kalman and Bertram (1960) are also of great value. Applications of Liapunov functions to process control are discussed in a review by Ellis and Storey (1963).

References

AIZERMAN, M. A. and F. R. GANTMACHER (1964), *Absolute stability of regulator systems*, Holden-Day, San Francisco.

ANDERSON, B. D. O. (1967), Application of the second method of Liapunov to the proof of the Markov stability criterion, *Int. J. Control*, **5,** 473–482.

BROCKETT, R. E. (1966), The status of stability theory for deterministic systems, *I.E.E.E. Trans. Aut. Control.* **AC–11,** 596–606.

CODDINGTON, E. A. and N. LEVINSON (1955), *Theory of ordinary differential equations*, McGraw-Hill, New York.

ELLIS, J. K. and C. STOREY (1963), Stability in process control: a review of the use of Liapunov functions, *Proc. Instn. Mech. Engrs.*, **178 Pt. 3 M,** 36–46.

GUREL, O. and L. LAPIDUS (1968), A guide to methods for the generation of Liapunov functions, IBM Tech. Dep. Rep. No. 320–2937.

HAHN, W. (1963), *Theory and application of Liapunov's direct method*, Prentice-Hall, Englewood Cliffs, N.J.

HARTMAN, P. (1964), *Ordinary differential equations*, John Wiley, New York.

HEWIT, J. (1968), A comparison of numerical methods in non-linear stability analysis, Ph.D. Thesis, Loughborough University of Technology.

HEWIT, J. R. and C. STOREY (1967), Optimization of the Zubov and Ingwerson methods for constructing Liapunov functions, *Electronics Letters*, **3,** No. 5, 211.

HEWIT, J. R. and C. STOREY (1969), Comparison of numerical methods in non-linear stability analysis, *Int. J. Control*, **10,** No. 6, 687–701.

INFANTE, E. F. and L. G. CLARKE (1964), A method for the determination of the domain of stability of second-order non-linear autonomous systems, *J. Appl. Mech.*, **31,** 315–320.

INFANTE, E. F. and L. G. CLARKE (1965), On the stability of the oscillations of a simple surge tank, *J. Appl. Mech.*, **32,** 945–947.

INGWERSON, D. R. (1961), A modified Liapunov method for non-linear stability analysis, *I.R.E. Trans. Aut. Control*, **AC–6,** 199–210.

KALMAN, R. E. and J. E. BERTRAM (1960), Control system analysis and design via the 'second method' of Liapunov: I. Continuous systems; II. Discrete-time systems, *Trans. A.S.M.E. J. Basic Eng.*, **82D,** 371–400.

KRASOVSKII, N. N. (1957), On stability with large initial perturbations, *P.M.M.*, **21,** 309–319.

KRASOVSKII, N. N. (1963), *Stability of motion, applications of Liapunov's second method to differential systems and equations with delay*, Stanford University Press, Stanford, Calif.

LA SALLE, J. P. and S. LEFSCHETZ (1961), *Stability by Liapunov's direct method with applications*, Academic Press, New York.

LEFSCHETZ, S. (1962), *Differential equations: geometric theory*, 2nd edn., John Wiley (Interscience), New York.

LEFSCHETZ, S. (1965), *Stability of non-linear control systems*, Academic Press, New York.

LEHNIGK, S. H. (1966), *Stability theorems for linear motions with an introduction to Liapunov's direct method*, Prentice-Hall, Englewood Cliffs, N.J.

LEHNIGK, S. H. (1967), Liapunov's direct method and the number of zeros with positive real parts of a polynomial with constant complex coefficients, *SIAM J. Control*, **5**, 234–244.

LIAPUNOV, A. M. (1907), Problème général de la stabilité du mouvement', *Annals of Math. Studies No. 7*, Princeton University Press, Princeton, N.J. (French translation of 1892 paper in Russian.)

LIAPUNOV, A. M. (1966), *Stability of motion*, Academic Press, New York.

MARDEN, M. (1966), *Geometry of polynominals*, American Math. Soc., Providence, Rhode Island.

PARKS, P. C. (1963), Analytic methods for investigating stability – linear and non-linear systems. A survey. *Proc. Instn. Mech. Engrs.*, **178 Pt. 3 M**, 7–17.

PARKS, P. C. (1969), Hermite-Hurwitz and Hermite-Bilharz links using matrix multiplication, *Electronics Letters*, **5**, 55.

RODDEN, J. J. (1964), Numerical applications of Liapunov stability theory, *J.A.C.C., Paper IX*–2, 261–268.

SCHULTZ, D. G. and J. E. GIBSON (1963), The variable gradient method for generating Liapunov functions, *Trans. A.I.E.E. Apps. Ind.*, **81**, 203–210.

SMITH, R. A. (1966), Matrix calculations for Liapunov quadratic forms, *J. Diff. Eqns.*, **2**, 208–217.

SZEGÖ, G. P. (1962), On a new partial differential equation for the stability analysis of time invariant control systems, *SIAM J. Control*, **1**, 63–75.

SZEGÖ, G. P. and G. R. GEISS (1963), A remark on 'A new partial differential equation for the stability analysis of time invariant control systems', *SIAM J. Control*, **1**, 369–376.

VOGT, W. G. 1964, Relative stability via the direct method of Liapunov. *Trans. A.S.M.E.*, **85D**, 87–90

WEATHERBURN, C. E. (1951), *Advanced vector analysis*, George Bell and Sons, Glasgow.

WILLEMS, J. L. (1970), *Stability theory of dynamical systems*, Nelson, London.

ZUBOV, V. I. (1962), *Mathematical methods for the study of automatic control systems*, Pergamon Press, Oxford.

ZUBOV, V. I. (1964), *Methods of A. M. Liapunov and their application*, Noordhoff, Groningen.

6

Solution of the Liapunov matrix equation

The whole of this chapter will be devoted to a study of various methods for solving the Liapunov matrix equation. Apart from its use in the stability analysis mentioned in the previous chapter the matrix equation also occurs in several other important problems including the evaluation of system functionals, certain aspects of non-linear stability analysis, optimal control theory, and matrix theory. Some of these other applications will be encountered later in this book and for a general survey with references we refer the reader to Barnett and Storey (1968).

6–1 Direct solution

We consider the equation

$$A'P + PA = -Q \tag{6–1–1}$$

for determining the symmetric matrix P, in which A is a real constant $n \times n$ matrix and Q is an arbitrary symmetric positive definite matrix. In order that (6–1–1) should have a unique solution it is assumed that A' and $-A$ have no common characteristic roots (see Theorem 2–10–2); that is, it is assumed that there are no roots λ_i, λ_j of A such that $\lambda_i + \lambda_j = 0$.

Since P and Q are symmetric Eqn (6–1–1) can be reduced to a set of $\frac{1}{2}n(n+1)$ simultaneous linear equations for the $\frac{1}{2}n(n+1)$ distinct elements of P. Thus the direct solution of (6–1–1) simply involves the inversion of a matrix of order $\frac{1}{2}n(n+1)$. We illustrate this for the simplest case of 2×2 matrices:

$$\begin{bmatrix} a_{11} & a_{21} \\ a_{12} & a_{22} \end{bmatrix} \begin{bmatrix} p_{11} & p_{12} \\ p_{12} & p_{22} \end{bmatrix} + \begin{bmatrix} p_{11} & p_{12} \\ p_{12} & p_{22} \end{bmatrix} \begin{bmatrix} a_{11} & a_{12} \\ a_{21} & a_{22} \end{bmatrix} = -\begin{bmatrix} q_{11} & q_{12} \\ q_{12} & q_{22} \end{bmatrix} \tag{6–1–2}$$

easily reduces to the three equations

$$\begin{aligned} 2a_{11}p_{11} + 2a_{21}p_{12} \qquad\qquad\qquad\quad &= -q_{11} \\ a_{12}p_{11} + (a_{11} + a_{22})p_{12} + a_{21}p_{22} &= -q_{12} \\ 2a_{12}p_{12} + 2a_{22}p_{22} \qquad\qquad\qquad\quad &= -q_{22}. \end{aligned} \tag{6–1–3}$$

In matrix form Eqn (6–1–3) can be written

$$Bp = -q \tag{6–1–4}$$

where
$$B = \begin{bmatrix} 2a_{11} & 2a_{21} & 0 \\ a_{12} & a_{11} + a_{22} & a_{21} \\ 0 & 2a_{12} & 2a_{22} \end{bmatrix},$$

$$p' = (p_{11}, p_{12}, p_{22}), \qquad q' = (q_{11}, q_{12}, q_{22}).$$

In the general $n \times n$ case the matrices p and q in (6–1–4) are given by

$$p' = (p_{11}, p_{12}, p_{22}, p_{13}, p_{23}, p_{33}, \ldots) \tag{6–1–5}$$

$$q' = (q_{11}, q_{12}, q_{22}, q_{13}, q_{23}, q_{33}, \ldots) \tag{6–1–6}$$

and B is formed from A as shown below

$$B = \begin{bmatrix} 2a_{11} & 2a_{21} & 0 & 2a_{31} & 0 & 0 & \cdots \\ a_{12} & a_{11} + a_{22} & a_{21} & a_{32} & a_{31} & 0 & \cdots \\ 0 & 2a_{12} & 2a_{22} & 0 & 2a_{32} & 0 & \cdots \\ a_{13} & a_{23} & 0 & a_{11} + a_{33} & a_{21} & a_{31} & \cdots \\ 0 & a_{13} & a_{23} & a_{12} & a_{22} + a_{33} & a_{32} & \cdots \\ 0 & 0 & 0 & 2a_{13} & 2a_{23} & 2a_{33} & \cdots \\ \cdot & \cdot & \cdot & \cdot & \cdot & \cdot & \cdots \\ \cdot & \cdot & \cdot & \cdot & \cdot & \cdot & \cdots \\ \cdot & \cdot & \cdot & \cdot & \cdot & \cdot & \cdots \end{bmatrix}. \tag{6–1–7}$$

An interesting systematic way of forming B from A which is suitable for use on a digital computer is given by Macfarlane (1963), and has been found to work well in practice.

Since, by assumption, the solution of (6–1–4) is unique it follows that B is non-singular and

$$p = -B^{-1}q. \tag{6–1–8}$$

A useful table of solutions P of (6–1–1) for A in companion form and a number of different forms of Q (up to fourth order) can be found in Ingwerson (1961).

6–2 Reduction by use of skew-symmetric matrix [Barnett and Storey (1966)]

It is obvious that the solution of (6–1–1) by means of (6–1–8) is a difficult computational problem for large values of n. We show in this section how the order of the matrix to be inverted can be reduced by n from $\frac{1}{2}n(n + 1)$ to $\frac{1}{2}n(n - 1)$ by the introduction of a certain real skew-symmetric matrix.

From
$$A'P + PA = -Q \tag{6–2–1}$$
we can write
$$(PA + \tfrac{1}{2}Q)' + (PA + \tfrac{1}{2}Q) = 0 \tag{6–2–2}$$
showing that
$$PA + \tfrac{1}{2}Q = S \tag{6–2–3}$$
where S is a skew-symmetric matrix which must satisfy
$$A'S + SA = \tfrac{1}{2}[A'Q - QA]. \tag{6–2–4}$$
This last condition arises directly from the fact that P is symmetric. Also, since A is non-singular (by assumption) it follows from (6–2–3) that
$$P = (S - \tfrac{1}{2}Q)A^{-1}. \tag{6–2–5}$$
Notice that in the stability analysis problem there is no need to calculate A^{-1}, since P is positive definite if and only if $A'PA$ is positive definite (Theorem 2–8–7).

Eqn (6–2–4) is of the same form as Eqn (6–2–1) and hence its solution can be written
$$s = \beta^{-1}\, d \tag{6–2–6}$$
where s, d, and β are formed from the elements of S, $\tfrac{1}{2}(A'Q - QA)$ and A respectively [cf. Eqn (6–1–8)]. Since, however, S is skew-symmetric the order of β is only $\tfrac{1}{2}n(n-1)$.

To obtain the matrix β from B in (6–1–7) the following procedure [Barnett (1967)] is adopted. Direct consideration of the elements of $A'S + SA$, i.e.,
$$[A'S + SA]_{ij} = \sum_{k=1}^{n} (s_{ik}a_{kj} - s_{jk}a_{ki})$$
shows that the rows and columns of B numbered 1, 3, 6, 10, . . ., $\tfrac{1}{2}n(n+1)$ have to be deleted. Also negative signs must be inserted symmetrically in the positions

row	column
1	3, 5, 8, 12, 17, 23 . . .
2, 3	6, 9, 13, 18, 24, . . .
4, 5, 6	10, 14, 19, 25, . . .
7, 8, 9, 10	15, 20, 26, . . .
. . . .	. . . .

together with the same arrangement but with 'rows' and 'columns' interchanged.

A rough estimation of the difference in computation involved in solving (6–2–1) directly by (6–1–8) or by (6–2–6) and (6–2–5) can be made as follows. Using Gauss's method the number of multiplications and divisions required to solve a system of n linear equations is $\frac{1}{3}n(n^2 + 3n - 1)$ [see, e.g., Faddeeva (1959)]. Here the reduction from $\frac{1}{2}n(n + 1)$ to $\frac{1}{2}n(n - 1)$ equations results in a saving of $(\frac{1}{4}n^5 + \frac{13}{12}n^3 - \frac{1}{3}n)$ operations. In using the second method, however, the following extra work must be carried out:

(*a*) the right-hand side of (6–2–4) must be calculated and this takes of the order of n^3 operations;

(*b*) the matrix A must be inverted for use in (6–2–5) at the cost of $\frac{1}{3}n(n^2 + 3n - 1)$ operations;

(*c*) P must now be computed from (6–2–5) at the cost of $\frac{1}{2}n^2(n + 1)$ operations.

It follows that the net saving resulting from the use of the skew-symmetric matrix is approximately $\frac{1}{4}n^5$ operations. This gives a relative reduction of roughly $\frac{1}{4}n^5 \Big/ \frac{1}{24}\, n^3(n + 1)^3 \approx \frac{6}{n}$, which is certainly a useful saving if n is not too large (< 50, say). In solving Eqn (6–2–6) on a digital computer the matrix β is easily obtained from A by a simple modification of Macfarlane's method. We also note that for small values of $n(< 5$, say) the skew-symmetric matrix technique enables the Liapunov matrix equation to be solved by hand calculation without too much effort.

It is also possible to use the facts that the matrix β is quite sparse (i.e., has a large number of zero elements) and that its zeros occur in a symmetrical manner to devise an iterative technique for the solution of (6–2–5). With this iteration method the largest matrix to be inverted has order $2n - 3$. The algebra involved in this latter technique is straightforward but tedious so we shall not reproduce it here. The reader is referred to Barnett (1967) for details.

We conclude this section with a simple 2×2 example.

Example 6–2–1

$$A = \begin{bmatrix} -1 & 0 \\ 0 & -2 \end{bmatrix} \qquad Q = \begin{bmatrix} 3 & 6 \\ 6 & 12 \end{bmatrix}$$

$$\tfrac{1}{2}[A'Q - QA] = \tfrac{3}{2}\left(\begin{bmatrix} -1 & 0 \\ 0 & -2 \end{bmatrix}\begin{bmatrix} 1 & 2 \\ 2 & 4 \end{bmatrix} - \begin{bmatrix} 1 & 2 \\ 2 & 4 \end{bmatrix}\begin{bmatrix} -1 & 0 \\ 0 & -2 \end{bmatrix}\right)$$

$$= \tfrac{3}{2}\begin{bmatrix} 0 & 2 \\ -2 & 0 \end{bmatrix}.$$

From (6–2–4)

$$\begin{bmatrix} -1 & 0 \\ 0 & -2 \end{bmatrix}\begin{bmatrix} 0 & s_{12} \\ -s_{12} & 0 \end{bmatrix} + \begin{bmatrix} 0 & s_{12} \\ -s_{12} & 0 \end{bmatrix}\begin{bmatrix} -1 & 0 \\ 0 & -2 \end{bmatrix} = \tfrac{3}{2}\begin{bmatrix} 0 & 2 \\ -2 & 0 \end{bmatrix}$$

giving $\qquad s_{12} = -1.$

But $$A^{-1} = \begin{bmatrix} -1 & 0 \\ 0 & -\frac{1}{2} \end{bmatrix}$$

so from (6–2–5)

$$P = \left(\begin{bmatrix} 0 & -1 \\ -1 & 0 \end{bmatrix} - \begin{bmatrix} \frac{3}{2} & 3 \\ 3 & 6 \end{bmatrix} \right) \begin{bmatrix} -1 & 0 \\ 0 & -\frac{1}{2} \end{bmatrix}$$

$$= \begin{bmatrix} \frac{3}{2} & 2 \\ 2 & 3 \end{bmatrix}$$

which can easily be checked by direct substitution in (6–2–1).

Exercise 6–2–1

Use the result of Exercise 2–8–2 to show that, under certain circumstances, solution of (6–2–1) can be reduced to finding a matrix Q which satisfies $A'Q = QA$.

Exercise 6–2–2

By replacing A by $A - \sigma I$ in (6–2–1) show that the real parts of the characteristic roots of A are less than σ if and only if the solution of

$$A'P + PA - 2\sigma P = -Q \qquad (6\text{–}2\text{–}7)$$

is positive definite [Kalman and Bertram (1960)].

Show that the solution of (6–2–7) can be written $P = (S_1 - \frac{1}{2}Q)(A - \sigma I)^{-1}$ where S_1 is the skew-symmetric solution of

$$A'S_1 + S_1A - 2\sigma S_1 = \tfrac{1}{2}(A'Q - QA).$$

6–3 Smith's method

Smith (1966) proved the following theorem:

Theorem 6–3–1

Let C be the companion matrix (Section 3–2) of the characteristic polynomial of the matrix A in (6–1–1) and let $K = \text{diag}\,(1, 0, 0, \ldots, 0)$. Then the solution of (6–1–1) is

$$P = \sum_i \sum_j u_{ij}(A')^{i-1}QA^{j-1} \qquad (6\text{–}3\text{–}1)$$

where $U = [u_{ij}]$ is the solution of

$$C'U + UC = -K. \qquad (6\text{–}3\text{–}2)$$

Smith gives an explicit but complicated expression for the solution U of (6–3–2). However, because of the simple forms of C and K it is possible to solve (6–3–2) in the following way [Barnett and Storey (1967*a*)]. As in the previous section the solution of (6–3–2) can be written

$$U = (S - \tfrac{1}{2}K)C^{-1} \tag{6–3–3}$$

where S is the skew-symmetric solution of

$$C'S + SC = \tfrac{1}{2}(C'K - KC). \tag{6–3–4}$$

Now the inverse of C is given by

$$C^{-1} = \begin{bmatrix} -c_{n-1}/c_n & -c_{n-2}/c_n & \cdots & -c_1/c_n & -1/c_n \\ 1 & 0 & \cdots & 0 & 0 \\ 0 & 1 & \cdots & 0 & 0 \\ \cdot & \cdot & \cdots & \cdot & \cdot \\ \cdot & \cdot & \cdots & \cdot & \cdot \\ \cdot & \cdot & \cdots & \cdot & \cdot \\ 0 & 0 & \cdots & 1 & 0 \end{bmatrix} \tag{6–3–5}$$

and the term on the right of (6–3–4) is simply

$$\tfrac{1}{2}(C'K - KC) = \tfrac{1}{2}\begin{bmatrix} 0 & -1 & 0 & \cdots & 0 \\ 1 & 0 & 0 & \cdots & 0 \\ 0 & 0 & 0 & \cdots & 0 \\ \cdot & \cdot & \cdot & \cdots & \cdot \\ \cdot & \cdot & \cdot & \cdots & \cdot \\ \cdot & \cdot & \cdot & \cdots & \cdot \\ 0 & 0 & 0 & \cdots & 0 \end{bmatrix}. \tag{6–3–6}$$

Thus Eqn (6–2–6) is in this case

$$\beta s = d \tag{6–3–7}$$

where

$$d' = (-\tfrac{1}{2}, 0, 0, \ldots, 0)$$

$$s' = (s_{12}, s_{13}, s_{23}, \ldots)$$

and the elements of the $\tfrac{1}{2}n(n-1) \times \tfrac{1}{2}n(n-1)$ matrix β are given by

$$\beta_{ij} = 1 \begin{cases} i = \tfrac{1}{2}J(J-1) + I; & j = \tfrac{1}{2}(J-1)(J-2) + I \\ I = 1, 2, \ldots, J-1; & J = 2, 3, \ldots, n-1 \\ i = j+1; & j = \tfrac{1}{2}(J-1)(J-2) + I \\ I = 1, 2, \ldots, J-2; & J = 3, 4, \ldots, n \end{cases} \tag{6–3–8}$$

$$\beta_{ij} = c_{n-p+1} \begin{cases} i = \frac{1}{2}(I-1)(I-2)+p \\ j = \frac{1}{2}(n-1)(n-2)+I \\ p = 1, 2, \ldots, I-1 \\ I = 1, 2, \ldots, n-1, \end{cases} \tag{6–3–9}$$

$$\beta_{ij} = -c_{n-r+1} \begin{cases} i = \frac{1}{2}(r-1)(r-2)+I \\ j = \frac{1}{2}(n-1)(n-2)+I \\ r = I+1, \ldots, n \\ I = 1, 2, \ldots, n-1, \end{cases} \tag{6–3–10}$$

and $\beta_{ij} = 0$, otherwise.

Notice that the matrix β is very sparse with no row having more than four non-zero elements. Thus solution of (6–3–2) using (6–3–3) and (6–3–4) should be quite straightforward. Having obtained U, Eqn (6–3–1) can now be used to find P. As mentioned earlier, it would also be possible to find P from (6–3–1) using the explicit formula for U given by Smith (1966). This, however, involves the calculation of the cofactors of the first row of the nth-order Hurwitz determinant. [See, e.g., Lehnigk (1966).]

Another related but more direct way of proceeding is to transform A into companion form (if possible) by means of a similarity transformation $C = HAH^{-1}$ (Section 3–2). This gives the following solution P for the matrix equation (6–1–1):

$$P = H'U_1H \tag{6–3–11}$$

where

$$Q = H'K_1H \tag{6–3–12}$$

and U_1 is the solution of

$$C'U_1 + U_1C = -K_1. \tag{6–3–13}$$

As before, U_1 can be written

$$U_1 = (S_1 - \tfrac{1}{2}K_1)C^{-1} \tag{6–3–14}$$

where S_1 is the skew-symmetric solution of

$$C'S_1 + S_1C = \tfrac{1}{2}(C'K_1 - K_1C). \tag{6–3–15}$$

This can be put into the form of Eqn (6–3–7) where s now is a vector of elements of S_1 and the matrix β is unchanged. Thus either when using the above direct method or Smith's equation (6–3–1) it is necessary to find the inverse of the matrix β. A method is available for partitioning β so that this inversion can be carried out easily. The algebra involved in the partitioning is straightforward but tedious [Barnett (1967)] so we shall not describe it in any detail here. In any case it will be seen that this method of solving the matrix equation is not as efficient as a method to be described later.

6–4 Solution when *A* is in Schwarz form

We recall from Chapter 3 (Section 3–4) that a matrix C in companion form can be changed into Schwarz form W by a similarity transformation

$$W = FCF^{-1}. \tag{6–4–1}$$

Thus a general (non-derogatory) matrix A can be converted first to companion form and then to Schwarz form by

$$W = (FH)A(FH)^{-1}. \tag{6–4–2}$$

It now follows that the solution of the matrix equation

$$A'P + PA = -Q \tag{6–4–3}$$

is given by

$$P = (FH)'U(FH) \tag{6–4–4}$$

with

$$Q = (FH)'D(FH) \tag{6–4–5}$$

where U is the solution of

$$W'U + UW = -D. \tag{6–4–6}$$

One reason for solving (6–4–3) by reducing A to Schwarz form lies in the fact that (6–4–6) can be solved explicitly for certain particular forms of D. Thus, for example, if

$$D = \text{diag}(0, 0, \ldots, 2w_n^2) \tag{6–4–7}$$

then $$U = \text{diag}(w_1 w_2 \ldots w_n, w_2 w_3 \ldots w_n, \ldots, w_{n-1}w_n, w_n) \tag{6–4–8}$$

[see Kalman and Bertram (1960)]. For this particular case the explicit solution of (6–4–3) with A in companion form C is given by $P = (p_{ij})$

with

$$\begin{aligned} p_{ij} &= 0, \quad i + j \text{ odd} \\ &= \sum_{k=r}^{i-1} (-1)^{k+i-1} c_{n-k}\, c_{n-i-j+1+k}, \quad i + j \text{ even} \end{aligned} \tag{6–4–9}$$

where

$$\begin{aligned} r &= 0 \text{ if } i + j \leqslant n + 1, \\ &= i + j - n - 1 \text{ if } i + j > n + 1. \end{aligned}$$

The corresponding right-hand side $Q = 2(q_{ij})$ of (6–4–3) is given by

$$\begin{aligned} q_{ij} &= 0, \quad n + i \text{ or } n + j \text{ odd} \\ &= c_{n+1-i}\, c_{n+1-j}, \quad n + i \text{ and } n + j \text{ even} \end{aligned} \tag{6–4–10}$$

with $c_0 = 1$, and $i, j = 1, 2, \ldots, n$. [See Ralston (1962) and Butchart (1965).]

Exercise 6–4–1 [Barnett (1967)]
Verify that if D in (6–4–6) is taken to be

$$D = \text{diag}\,(0, 0, \ldots, 2w_n w_{n-1}, 0)$$

then the solution of (6–4–6) is given by $U = (u_{ij})$

with

$$u_{nn} = 1$$

$$u_{n-1,n-1} = w_n^2 + w_{n-1} + w_{n-2}$$

$$u_{ii} = w_i w_{i+1} \ldots w_{n-2}(w_n^2 + w_i + w_{i-1}), \quad i = 1, 2, \ldots, n-2 \; (w_0 = 0)$$

$$u_{i,i+2} = -w_i w_{i+1} \ldots w_{n-2}, \quad i = 1, 2, \ldots, n-2$$

$$u_{n-1,n} = w_n$$

and all other $u_{ij} = 0$.

Exercise 6–4–2
Give a similar result for D given by (d_{ij}) where $d_{kk} = 2w_k w_{k+1} \cdots w_n$ (with k fixed) all other $d_{ij} = 0$.

We illustrate the use of the Schwarz form by carrying out the calculations in full for the case of a 3×3 matrix.

Example 6–4–1
Start with

$$A = \begin{bmatrix} 2 & -6 & -2 \\ -2 & 1 & 1 \\ 6 & -9 & -4 \end{bmatrix}$$

and reduce it to companion form by the similarity transformation HAH^{-1} with

$$H^{-1} = \begin{bmatrix} 1 & 1 & 1 \\ 0 & -1 & 0 \\ 2 & 3 & 1 \end{bmatrix}$$

$$H = \begin{bmatrix} -1 & 2 & 1 \\ 0 & -1 & 0 \\ 2 & -1 & -1 \end{bmatrix}$$

the companion form being

$$C = \begin{bmatrix} 0 & 1 & 0 \\ 0 & 0 & 1 \\ -2 & 1 & -1 \end{bmatrix}.$$

(The characteristic equation of A is just $\lambda^3 + \lambda^2 - \lambda + 2 = 0$.)

Now reduce C to Schwarz form by the similarity transformation FCF^{-1} with

$$F = \begin{bmatrix} 1 & 0 & 0 \\ 0 & 1 & 0 \\ 2 & 0 & 1 \end{bmatrix}$$

$$F^{-1} = \begin{bmatrix} 1 & 0 & 0 \\ 0 & 1 & 0 \\ -2 & 0 & 1 \end{bmatrix}$$

the Schwarz form being

$$W = \begin{bmatrix} 0 & 1 & 0 \\ -2 & 0 & 1 \\ 0 & 3 & -1 \end{bmatrix}.$$

The solution of $W'U + UW = -\text{diag}\ (0, 0, 2)$ is [from (6–4–8)]

$$U = \text{diag}\ (-6, -3, 1).$$

The corresponding solution of $C'P + PC = -Q$ is given by $P = (p_{ij})$ with [from (6–4–9)]

$$\begin{aligned} p_{12} &= p_{13} = 0 \\ p_{11} &= c_2c_3 = -2 \\ p_{22} &= -c_3 + c_2c_1 = -3 \\ p_{33} &= 1 \\ p_{13} &= c_3 = 2, \end{aligned}$$

i.e.

$$P = \begin{bmatrix} -2 & 0 & 2 \\ 0 & -3 & 0 \\ 2 & 0 & 1 \end{bmatrix}.$$

Similarly [from (6–4–10)] the matrix $Q = 2(q_{ij})$ is given by

$$\begin{aligned} q_{12} &= q_{21} = q_{22} = q_{23} = q_{32} = 0 \\ q_{11} &= c_3^2 = 4 \\ q_{31} &= q_{13} = c_3c_1 = 2 \\ q_{33} &= c_1^2 = 1 \end{aligned}$$

so that

$$Q = \begin{bmatrix} 8 & 0 & 4 \\ 0 & 0 & 0 \\ 4 & 0 & 2 \end{bmatrix}.$$

Finally, the solution of (6–4–3) is given by $P = (FH)' \operatorname{diag}(-6, -3, 1)(FH)$ with $Q = (FH)' \operatorname{diag}(0, 0, 2)(FH)$. Now

$$FH = \begin{bmatrix} 1 & 0 & 0 \\ 0 & 1 & 0 \\ 2 & 0 & 1 \end{bmatrix} \begin{bmatrix} -1 & 2 & 1 \\ 0 & -1 & 0 \\ 2 & -1 & -1 \end{bmatrix}$$

$$= \begin{bmatrix} -1 & 2 & 1 \\ 0 & -1 & 0 \\ 0 & 3 & 1 \end{bmatrix}$$

so that

$$P = \begin{bmatrix} -6 & 12 & 6 \\ 12 & -18 & -9 \\ 6 & -9 & -5 \end{bmatrix}$$

and

$$Q = \begin{bmatrix} 0 & 0 & 0 \\ 0 & 18 & 6 \\ 0 & 6 & 2 \end{bmatrix}.$$

Another reason for transforming the matrix A into Schwarz form is that the matrix β in (6–2–6) can then be made triangular by an appropriate choice of the vectors s and d. [Solution of (6–2–6) by inversion of β is then almost trivial – see Exercise 2–9–1.] Once again the algebra involved, although straightforward, is extremely tedious and we illustrate the technique for the 3×3 case. [For details of the scheme for general $n \times n$ matrices see Barnett (1967); see also Parks (1963).]

The solution of

$$W'P + PW = -D \tag{6–4–11}$$

is required with

$$W = \begin{bmatrix} 0 & 1 & 0 \\ -w_1 & 0 & 1 \\ 0 & -w_2 & -w_3 \end{bmatrix}. \tag{6–4–12}$$

Following Section 6–2 the solution of (6–4–11) can be written

$$P = (S - \tfrac{1}{2}D)W^{-1} \tag{6–4–13}$$

where S is the skew-symmetric solution of

$$W'S + SW = \tfrac{1}{2}(W'D - DW). \tag{6–4–14}$$

Once S is known the solution of (6–4–11) is easily obtained from (6–4–13) since W^{-1} is simply (Section 3–5)

$$W^{-1} = \begin{bmatrix} -\frac{w_2}{w_1 w_3} & -\frac{1}{w_1} & -\frac{1}{w_1 w_3} \\ 1 & 0 & 0 \\ -\frac{w_2}{w_3} & 0 & -\frac{1}{w_3} \end{bmatrix}. \tag{6–4–15}$$

Eqn (6–4–14) written in full is

$$\begin{bmatrix} 0 & -w_1 & 0 \\ 1 & 0 & -w_2 \\ 0 & 1 & -w_3 \end{bmatrix} \begin{bmatrix} 0 & s_{12} & s_{13} \\ -s_{12} & 0 & s_{23} \\ -s_{13} & -s_{23} & 0 \end{bmatrix} + \begin{bmatrix} 0 & s_{12} & s_{13} \\ -s_{12} & 0 & s_{23} \\ -s_{13} & -s_{23} & 0 \end{bmatrix} \begin{bmatrix} 0 & 1 & 0 \\ -w_1 & 0 & 1 \\ 0 & -w_2 & -w_3 \end{bmatrix}$$

$$= \tfrac{1}{2}\left\{ \begin{bmatrix} 0 & -w_1 & 0 \\ 1 & 0 & -w_2 \\ 0 & 1 & -w_3 \end{bmatrix} \begin{bmatrix} d_{11} & d_{12} & d_{13} \\ d_{12} & d_{22} & d_{23} \\ d_{13} & d_{23} & d_{33} \end{bmatrix} - \begin{bmatrix} d_{11} & d_{12} & d_{13} \\ d_{12} & d_{22} & d_{23} \\ d_{13} & d_{23} & d_{33} \end{bmatrix} \begin{bmatrix} 0 & 1 & 0 \\ -w_1 & 0 & 1 \\ 0 & -w_2 & -w_3 \end{bmatrix} \right\} \tag{6–4–16}$$

so that (6–2–6) becomes

$$\begin{bmatrix} 0 & -w_2 & 0 \\ 1 & -w_3 & -w_1 \\ 0 & 1 & -w_3 \end{bmatrix} \begin{bmatrix} s_{12} \\ s_{13} \\ s_{23} \end{bmatrix} = \tfrac{1}{2} \begin{bmatrix} -w_1 d_{22} - d_{11} + d_{13} w_2 \\ -w_1 d_{23} - d_{12} + w_3 d_{13} \\ d_{13} - w_2 d_{33} - d_{22} + w_3 d_{23} \end{bmatrix}. \tag{6–4–17}$$

This is easily rearranged to give

$$\begin{bmatrix} -w_2 & 0 & 0 \\ 1 & -w_3 & 0 \\ -w_3 & -w_1 & 1 \end{bmatrix} \begin{bmatrix} s_{13} \\ s_{23} \\ s_{12} \end{bmatrix} = \tfrac{1}{2} \begin{bmatrix} -w_1 d_{22} + d_{11} - d_{13} w_2 \\ d_{13} - w_2 d_{23} - d_{22} + w_3 d_{23} \\ -w_1 d_{23} - d_{12} + w_3 d_{13} \end{bmatrix} \tag{6–4–18}$$

which is in lower triangular form. It is worth pointing out also that in many cases when A in Eqn (6–4–3) is tridiagonal (thus including the Schwarz form as a particular instance) then diagonal matrices P and Q satisfying Eqn (6–4–3) can be found (see Section 7–3).

Exercise 6–4–3

Use Eqns (6–4–7) and (6–4–8) to show that once the matrix A has been transformed into Schwarz form the stability problem is solved immediately; a necessary and sufficient condition for asymptotic stability being that all w_i are positive.

6–5 Infinite series solution

When the matrix A is a stability matrix then

$$\alpha = (I + A')(I - A')^{-1} \tag{6–5–1}$$

is convergent [Taussky (1964)]. From (6–5–1) it is seen that

$$A' = (\alpha + I)^{-1}(\alpha - I) \tag{6–5–2}$$

so that for this particular A' the matrix equation

$$A'P + PA = -Q \tag{6–5–3}$$

becomes

$$P - \alpha P \alpha' = \tfrac{1}{2}(\alpha + I)Q(\alpha' + I). \tag{6–5–4}$$

If we denote the right side of (6–5–4) by M then since

$$\alpha + I = 2(I - A')^{-1}$$

it follows that M is given in terms of A by

$$M = 2(I - A')^{-1}Q(I - A)^{-1}. \tag{6–5–5}$$

Now (6–5–4) can be written [Wedderburn (1904)]

$$(I - \phi)P = M \tag{6–5–6}$$

and hence the solution P of (6–5–3) is given by the infinite series

$$\begin{aligned} P &= (I - \phi)^{-1}M. \\ &= (I + \phi + \phi^2 + \phi^3 + \cdots)M \\ &= M + \alpha M \alpha' + \alpha^2 M(\alpha')^2 + \cdots \end{aligned} \tag{6–5–7}$$

Under the assumption that A is a stability matrix we now prove that the series (6–5–7) converges.

When A is a stability matrix α is convergent and $\alpha^n \to 0$; hence the characteristic roots $\lambda_i(\alpha)$ of α are less than one in modulus (Section 2–9). If the λ_i are distinct then α can be diagonalized by a similarity transformation

$$L\alpha L^{-1} = \Lambda = \text{diag}(\lambda_1, \lambda_2, \ldots, \lambda_n) \tag{6–5–8}$$

from which it follows that

$$\alpha = L^{-1}\Lambda L, \; \alpha' = L'\Lambda(L^{-1})'$$

and

$$\alpha^r M (\alpha')^r = L^{-1}\Lambda^r L M L' \Lambda^r (L^{-1})', \; r = 1, 2, 3, \ldots \tag{6–5–9}$$

Putting $LML' = K \equiv (k_{ij})$ in (6–5–9) and using the result in (6–5–7) gives

$$P = L^{-1}[K + \Lambda K \Lambda + \Lambda^2 K \Lambda^2 + \cdots](L^{-1})'. \tag{6–5–10}$$

Since $(\Lambda^r K \Lambda^r)_{ij} = k_{ij} \lambda_i^r \lambda_j^r$, the ijth element of the matrix in square brackets in (6–5–10) is just $k_{ij}(1 - \lambda_i \lambda_j)^{-1}$ which converges since $|\lambda_i \lambda_j| = |\lambda_i||\lambda_j| < 1$. Hence the infinite matrix series (6–5–7) for P also converges. [A more general convergence proof which does not require the λ_i to be distinct can be found in Redheffer (1965).]

If in (6–5–3) Q is taken equal to $2(\alpha + I)^{-1}(\alpha' + I)^{-1}$ it follows from (6–5–4) that M is the unit matrix and the solution $P = T$ of (6–5–3) is then

$$T = I + \alpha\alpha' + \alpha^2(\alpha')^2 + \alpha^3(\alpha')^3 + \cdots \tag{6–5–11}$$

If, on the other hand, Q in (6–5–3) is I then

$$M = \tfrac{1}{2}(I + \alpha + \alpha' + \alpha\alpha')$$

and from (6–5–7)

$$\begin{aligned} 2P &= (I + \alpha + \alpha' + \alpha\alpha') + (\alpha\alpha' + \alpha^2\alpha' + \alpha(\alpha')^2 + \alpha^2(\alpha')^2) + \cdots \\ &= (I + \alpha\alpha' + \alpha^2(\alpha')^2 + \cdots) + (\alpha\alpha' + \alpha^2(\alpha')^2 + \cdots) + \\ &\qquad \alpha(I + \alpha\alpha' + \cdots) + (I + \alpha\alpha' + \cdots)\alpha' \\ &= T + (T - I) + \alpha T + T\alpha'. \end{aligned}$$

Thus when $Q = I$ in (6–5–3) the solution is

$$P = T - \tfrac{1}{2}I + \tfrac{1}{2}(\alpha T + (\alpha T)') \tag{6–5–12}$$

with T given by (6–5–11).

The use of the infinite series solution of (6–5–3) as a computational technique depends, of course, on its rate of convergence. If μ_i is a characteristic root of A then the corresponding root of α is

$$\lambda_i = \frac{1 + \mu_i}{1 - \mu_i}. \tag{6–5–13}$$

This shows that convergence will be most rapid when $\mu_i = -1$ and slowest when $|\mu_i|$ is very large or very small. The calculation of each matrix in the series will require $2n^3$ multiplications and the calculation of A^{-1} and $(I - A)^{-1}$ will also require order n^3 operations. Thus if the series is fairly quickly convergent it will be useful from a computational point of view.

Practical experience of the use of the infinite series (6–5–7) for computing the solution of (6–5–3) indicates that when used directly the convergence is much too slow [for details of an 8×8 example see Barnett (1967)]. A great improvement is achieved by the following 'nesting' procedure devised by Smith (1968).

To obtain the partial sums of

$$T = I + \alpha\alpha' + \alpha^2(\alpha')^2 + \alpha^3(\alpha')^3 + \cdots$$

Smith suggests setting

$$\begin{aligned} T_1 &= I \\ T_2 &= T_1 + \alpha T_1 \alpha' \\ T_3 &= T_2 + \alpha^2 T_2 (\alpha')^2 \\ &\;\;\vdots \\ T_{k+1} &= T_k + \alpha^{2^{k-1}} T_k (\alpha')^{2^{k-1}} \end{aligned}$$

so that $T_{k+1} = I + \alpha\alpha' + \alpha^2(\alpha')^2 + \cdots + \alpha^{(2^k)-1}(\alpha')^{(2^k)-1}$.

By this simple device over 4,000 terms of the series are obtained in only twelve iterations.

The transformation (6–5–1) also occurs in the theory of linear discrete time systems (see Section 5–4) of the form

$$x(t_{k+1}) = \alpha' x(t_k) \qquad (6\text{–}5\text{–}14)$$

since the quadratic form $V = x'Px$ is a Liapunov function for (6–5–14) if and only if the symmetric solution P of

$$P - \alpha P \alpha' = M \qquad (6\text{–}5\text{–}15)$$

for an arbitrary symmetric positive definite matrix M, is itself positive definite. [See, e.g., Kalman and Bertram (1960).] The reader is also referred to Smith (1966) for a generalization to cases where the elements of A are complex and to Zubov (1959) for the use of (6–5–1) to obtain results on approximate regions in parameter space for stability of continuous linear systems.

Finally, we note that the method of this section can also be used to solve the more general matrix equation

$$AX + XB = C. \qquad (6\text{–}5\text{–}16)$$

Exercise 6–5–1

Write down the nesting scheme for the series (6–5–7).

6–6 Use of Kronecker product [Barnett and Storey (1967*b*)]

The matrix equation

$$A'P + PA = -Q \qquad (6\text{–}6\text{–}1)$$

can be written

$$[(A' \otimes I_n) + (I_n \otimes A')]\rho = \phi \qquad (6\text{–}6\text{–}2)$$

where ρ and ϕ are column vectors formed by taking the rows of P and $-Q$ respectively. (See Theorem 2–10–2.) Thus it is seen from (6–6–2) that

$$\rho = [(A' \otimes I_n) + (I_n \otimes A')]^{-1}\phi \qquad (6\text{–}6\text{–}3)$$

gives the elements of P, the solution of (6–6–1).

We now consider the inverse of the matrix in square brackets in (6–6–3). If this matrix is denoted by R then it is rather more convenient to work with its transpose R',

$$R' = \begin{bmatrix} a_{11}I + A & a_{12}I & a_{13}I & \cdots & \cdot \\ a_{21}I & a_{22}I + A & a_{23}I & \cdots & \cdot \\ & \cdot & & & \cdot \\ & & \cdot & & \cdot \\ & & & \cdot & \cdot \\ & & & & \cdot \\ & & & & \cdot \\ & & & & a_{nn}I + A \end{bmatrix}. \tag{6–6–4}$$

First notice that all the submatrices of R' in (6–6–4) commute with each other so that the inversion of R' involves exactly the same operations as the inversion of the matrix obtained by replacing I and A by the scalars 1 and a respectively. This latter matrix can be written $A + I_n a$ showing the close connection between the inversion of R' and the characteristic matrix $\lambda I_n - A$.

From Theorem 2–2–2 we have

$$(\lambda I_n - A)^{-1} = B(\lambda)/\,|\,\lambda I_n - A\,| \tag{6–6–5}$$

where
$$|\,\lambda I_n - A\,| = \lambda^n + c_1\lambda^{n-1} + c_2\lambda^{n-2} + \cdots + c_n \tag{6–6–6}$$

and
$$B(\lambda) = \lambda^{n-1}I_n + B_1\lambda^{n-2} + B_2\lambda^{n-3} + \cdots + B_{n-1} \tag{6–6–7}$$

with
$$B_1 = A + c_1 I_n \tag{6–6–8}$$
$$B_k = AB_{k-1} + c_k I_n.$$

It now follows by taking $\lambda = -a$ in (6–6–5), (6–6–7), and (6–6–8) that

$$(R')^{-1} = (I_n \otimes D^{-1})(I_n \otimes A^{n-1} - B_1 \otimes A^{n-2} + B_2 \otimes A^{n-3} - \cdots \\ \cdots - (-1)^n B_{n-1} \otimes I_n) \tag{6–6–9}$$

where
$$D = A^n - c_1A^{n-1} + c_2A^{n-2} \cdots + (-1)^n c_n I_n. \tag{6–6–10}$$

Furthermore, since A satisfies its own characteristic equation (Theorem 2–2–6) D can be written

$$-2(c_1A^{n-1} + c_3A^{n-3} + \cdots + c_{n-1}A), \text{ for } n \text{ even} \tag{6–6–11}$$

and
$$-2(c_1A^{n-1} + c_3A^{n-3} + \cdots + c_nI_n), \text{ for } n \text{ odd.} \tag{6–6–12}$$

Notice that the only matrix to be inverted is D which is $n \times n$.

Exercise 6–6–1
Find the inverse of the matrix

$$U = A \otimes I_n + I_n \otimes C$$

where C is a square matrix of the same order as A. How can this be used to solve the matrix equation

$$AX + XC' = Q?$$

6–7 Finite series method [Bickley and MacNamee (1960) and Jameson (1968)]

Again we start with

$$A'P + PA = -Q \tag{6–7–1}$$

and assume that the characteristic equation of A is

$$|\lambda I_n - A| = \lambda^n + c_1\lambda^{n-1} + c_2\lambda^{n-2} + \cdots + c_n = 0$$

so that, by the Cayley-Hamilton theorem (Theorem 2–2–6), it follows that

$$A^n + c_1A^{n-1} + c_2A^{n-2} + \cdots + c_nI = 0. \tag{6–7–2}$$

Define the sequence of matrices

$$\begin{aligned}
Q_0 &= 0\\
Q_1 &= -Q = A'P + PA\\
Q_2 &= A'Q_1 - Q_1A + A'Q_0A = (A')^2P - PA^2\\
Q_3 &= A'Q_2 - Q_2A + A'Q_1A = (A')^3P + PA^3\\
&\cdots\cdots\cdots\cdots\cdots\cdots\cdots\cdots\\
Q_k &= A'Q_{k-1} - Q_{k-1}A + A'Q_{k-2}A = (A')^kP - (-1)^kPA^k\\
&\cdots\cdots\cdots\cdots\cdots\cdots\cdots\cdots
\end{aligned}$$

then

$$\begin{aligned}
&Q_n - c_1Q_{n-1} + \cdots + (-1)^{n-1}c_{n-1}Q_1\\
&= (A')^nP - (-1)^nPA^n - c_1[(A')^{n-1}P - (-1)^{n-1}PA^{n-1}] +\\
&\qquad \cdots + (-1)^{n-1}c_{n-1}(A'P + PA)\\
&= (A')^nP - c_1(A')^{n-1}P + \cdots + (-1)^{n-1}c_{n-1}A'P\\
&\qquad + (-1)^{n-1}P\{A^n + c_1A^{n-1} + \cdots + c_{n-1}A\}\\
&= \{(A')^n - c_1(A')^{n-1} + \cdots + (-1)^nc_nI\}P
\end{aligned}$$

on using (6–7–2).

Thus we have

$$P = G^{-1}(Q_n - c_1Q_{n-1} + \cdots + (-1)^{n-1}c_{n-1}Q) \tag{6–7–3}$$

where $G = (A')^n - c_1(A')^{n-1} + \cdots + (-1)^nc_nI.$

If it is more convenient the transpose of (6–7–3) can be taken to give

$$P = (-1)^{n-1}(Q_n + c_1 Q_{n-1} + \cdots + c_{n-1}Q_1)(G')^{-1} \qquad (6\text{–}7\text{–}4)$$

with

$$G' = A^n - c_1 A^{n-1} + \cdots (-1)^n c_n I. \qquad (6\text{–}7\text{–}5)$$

Notice that G' is identical to the matrix D of (6–6–10).

The Cayley-Hamilton theorem can again be used to simplify Eqn (6–7–5) as in (6–6–11) or (6–6–12).

The greatest disadvantage of this method from the computational point of view is that the coefficients of the characteristic polynomial have to be calculated. To offset this disadvantage, however, it is only necessary to invert an $n \times n$ matrix. Jameson (1968) carries out the above analysis for the more general matrix equation $AX + XB = C$. (It is interesting that the method proposed by Jameson was given much earlier in the little known paper by Bickley and MacNamee, 1960.)

Exercise 6–7–1 [Jameson (1968)]
Show that the explicit solution of (6–7–1) for a 2×2 matrix is given by Eqn (6–7–4) as

$$P = \frac{-1}{2 \operatorname{tr}(A)} \{Q + |A| (A^{-1})' Q (A^{-1})\}.$$

(Use Theorem 2–2–1.)

Discussion and further reading

It is not possible in this book to discuss the numerical analysis aspects of the work. On the basis of somewhat limited experience, however [Barnett (1967)], it appears that the most promising technique for matrices of order > 10 is the infinite series with Smith's iterative scheme. The direct method breaks down if A is ill-conditioned or the number of equations $[\frac{1}{2}n(n+1)]$ is large. The reduction to $\frac{1}{2}n(n-1)$ equations achieved by introducing the skew-symmetric matrix S can help in some cases but even with this method the capacity of the largest computers is soon exceeded. The use of the companion or Schwarz forms, although attractive from a theoretical point of view, suffers from the difficulties that arise in transforming a general matrix into companion or Schwarz form [see Wilkinson (1965)]. The Kronecker product method and Jameson's method only involve inverting a matrix of order n but also require calculation of the coefficients of the characteristic polynomial of A.

We have not tried to be exhaustive in this chapter either in our treatment of the methods of solution of the matrix equation or of our coverage of references to related work in the field. The reader is, however, referred to the work of Power [1967(*a*), 1967(*b*), 1967(*c*), 1968(*a*), 1968(*b*), and 1969] and to the paper by Frame (1963). Of the references we list both Smith and Jameson give many further useful references.

References

Barnett, S. and C. Storey (1966), Stability analysis of constant linear systems by Liapunov's second method, *Electronics Letters*, **2,** 165–166.

Barnett, S. (1967), Solution of some algebraic problems arising in the theory of stability and sensitivity of systems, with particular reference to the Liapunov matrix equation, Ph.D. Thesis, Loughborough University of Technology.

Barnett, S. and C. Storey (1967*a*) Analysis and synthesis of stability matrices, *J. Diff. Eqns.*, **3,** 414–422.

Barnett, S. and C. Storey (1967*b*) On the general functional-matrix for a linear system, *I.E.E.E. Trans. Aut. Control* (Short Papers), **AC–12,** 436–438.

Barnett, S. and C. Storey (1968), Some applications of the Liapunov matrix equation, *J. Inst. Math. App.*, **4,** 33–42.

Bickley, W. G. and J. MacNamee (1960), Matrix and other direct methods for the solution of systems of linear difference equations, *Phil. Trans. Roy. Soc. Lond.*, Series A, **252,** 69–131.

Butchart, R. L. (1965), An explicit solution to the Fokker-Plank equation for an ordinary differential equation, *Int. J. Control*, **1,** 201–208.

Faddeeva, V. N. (1959), *Computational methods of linear algebra*, Dover, New York.

Frame, J. S. (1963), The rectangular matrix equation $AY + B = YC$, *Notices Am. Math. Soc.*, **10,** 566.

Ingwerson, D. R. (1961), A modified Liapunov method for non-linear stability analysis, *I.R.E. Trans. Aut. Control*, **AC–6,** 199–210.

Jameson, A. (1968), Solution of the equation $AX + XB = C$ by the inversion of an $M \times M$ or $N \times N$ matrix, *SIAM J. Appl. Math.*, **16,** 1020–1023.

Kalman, R. E. and J. E. Bertram (1960), Control system analysis and design via the 'second method' of Liapunov: I. Continuous systems; II. Discrete-time systems, *Trans. A.S.M.E. J. Basic Engng.*, **82D,** 371–400.

Lehnigk, S. H. (1966), *Stability theorems for linear motions* with an introduction to Liapunov's direct method, Prentice-Hall, Englewood Cliffs, N.J.

Macfarlane, A. G. J. (1963), The calculation of functionals of the time and frequency response of a linear constant coefficient dynamical system, *Q.J. Mech. Appl. Math.*, **16,** 259–271.

Parks, P. C. (1963), A new proof of the Hurwitz stability criterion by the second method of Liapunov with applications to 'optimum' transfer functions, *Preprints Joint Automatic Control Conf.*

Power, H. M. (1967*a*), The companion matrix and Liapunov functions for linear, multivariable, time-invariant systems, *J. Franklin Inst.*, **283,** 214–234.

Power, H. M. (1967*b*), Further comments on the Liapunov matrix equation, *Electronics Letters*, **3,** 153–154.

Power, H. M. (1967*c*), Solution of the Liapunov matrix equation with a diagonal input matrix, obtained without matrix inversion, *Electronics Letters*, **3,** 325–326.

Power, H. M. (1968*a*), Liapunov matrix equations with positive semi-definite input matrices, *Electronics Letters*, **4,** 325–326.

Power, H. M. (1968*b*), Two applications of the matrix transformation $A = (B + I)(B - I)^{-1}$ in dynamical system theory, *Electronics Letters*, **4,** 479–481.

Power, H. M. (1969), Solution of the Liapunov matrix equation with a diagonal input matrix, obtained without matrix inversion, *Electronics Letters*, **5,** 135–136.

Ralston, A. (1962), A symmetric matrix formulation of the Hurwitz-Routh stability criterion, *Trans. I.R.E. (Correspondence)*, **AC–7,** 50–51.

Redheffer, R. (1965), Remarks on a paper of Taussky, *J. Algebra*, **2,** 42–47.

SMITH, R. A. (1966), Matrix calculations for Liapunov quadratic forms, *J. Diff. Eqns.*, **2,** 208–217.

SMITH, R. A. (1968), Matrix equation $XA + BX = C$, *SIAM J. Appl. Math.*, **16,** 198–201.

TAUSSKY, O. (1964), On the variation of the characteristic roots of a finite matrix under various changes of its elements, *Recent Advances in Matrix Theory*, ed. H. Schneider, University of Wisconsin Press, 125–138.

WEDDERBURN, M. (1904), Note on the linear matrix equation, *Proc. Edin. Math. Soc.*, **22,** 49–53.

WILKINSON, J. H. (1965), *The algebraic eigenvalue problem*, Oxford University Press, London.

ZUBOV, V. I. (1959), A new method of constructing stability regions for an automatic control system in the space occupied by the permissible values of the parameters, *Automat. Rem. Control*, **20,** 314–318.

7

Some properties of stability matrices

7–1 General expression for a stability matrix

Using the Liapunov matrix equation (5–3–3) a general expression for a stability matrix is very easily written down:

Theorem 7–1–1 [*Barnett and Storey* (*1967*)]
The matrix

$$A = P^{-1}(S - Q), \tag{7–1–1}$$

where P, Q are arbitrary symmetric positive definite matrices and S is an arbitrary skew-symmetric matrix is always a stable matrix.

Proof. Since $A'P + PA = -2Q$, A is stable by Theorem 5–3–2.

Further, since P is positive definite so is P^{-1} (Theorem 2–8–1) so (7–1–1) may be written

$$A = P_1(S - Q) \tag{7–1–2}$$

where P_1 is arbitrary symmetric positive definite. Another alternative expression to (7–1–1) or (7–1–2) is

$$A = (S - Q)P, \tag{7–1–3}$$

for in this case $A'P + PA = -2PQP$, and PQP is positive definite by virtue of Theorem 2–8–7. It should be noticed that the converse also holds: given any stable matrix A it can be expressed in the form (7–1–1), the matrices P and S then being uniquely determined by A through Eqns (6–2–1) and (6–2–4).

If A in (7–1–1) or (7–1–3) is the matrix of a system of linear differential equations

$$\dot{x} = Ax \tag{7–1–4}$$

then $V = x'Px$ is a Liapunov function with $dV/dt = -2x'Qx$ for (7–1–1) and $dV/dt = -2x'PQPx$ for (7–1–3). In either case the ratio $(dV/dt)/V$ is independent of S, so that the 'speed' at which the solution of (7–1–4) approaches the origin is independent of S (Theorem 5–2–9). This implies that the magnitudes of the real parts of the characteristic roots of A are essentially independent of S, for the solution of (7–1–4) is $x(t) = \exp(At)x(0)$ (Exercise 4–2–4). This can be demonstrated more precisely by using a result due to Vogt (1965). Let α_i, β_i be the real and imaginary parts respectively of the characteristic root $\lambda_i(A)$, numbered so that $0 > \alpha_1 \geqslant \alpha_2 \geqslant \ldots \geqslant \alpha_n$ and let μ_i be the characteristic roots of $P^{-1}Q$. Since P, Q are both positive definite it follows from Theorem 2–8–8 that the μ_i are all real and positive, so that they may be ordered as $\mu_n \geqslant \cdots \geqslant \mu_2 \geqslant \mu_1 > 0$. Then by Vogt's theorem it follows that:

Theorem 7–1–2

$$\alpha_1 \leqslant -\mu_1 \text{ and } \alpha_n \geqslant -\mu_n. \tag{7–1–5}$$

That is, the real parts of the characteristic roots of A in (7–1–1) all lie between the greatest and least roots of the matrix $-P^{-1}Q$. A similar result also applies to (7–1–3), the μ_i then being the roots of QP.

For example, taking $P_1 = M$, an arbitrary positive definite matrix, and $Q = I$ in (7–1–2) gives a stability matrix $A_1 = MS_1 - M$, and $P_1 = I$, $Q = M$ gives $A_2 = S_2 - M$. Since the matrix product P_1Q is the same in both cases it follows from Theorem 7–1–1 that the real parts of the characteristic roots of A_1 and A_2 all lie between the greatest and least characteristic roots of the negative definite matrix $-M$ for any skew-symmetric matrices S_1 and S_2, although A_1 and A_2 are in general of quite different form. The expressions for A_1 and A_2 can perhaps be thought of as generalizations of the fact that any negative definite matrix is a stability matrix.

The role of the skew-symmetric matrix S in (7–1–1) can be clarified a little. Certainly the characteristic roots of A are all real if $S = 0$, but they are not necessarily complex if $S \neq 0$. However, the roots of the matrix $P^{-1}S$ are either zero or pure imaginary (Exercise 2–8–3). Let them be denoted by $\pm j\gamma_1, \pm j\gamma_2, \ldots, \pm j\gamma_m$ where j denotes $(-1)^{1/2}$, $\gamma_1 \geqslant \gamma_2 \geqslant \ldots \geqslant \gamma_m > 0$ and $m \leqslant n/2$. By an argument similar to that used by Vogt (1965) to obtain (7–1–5) it is now shown that:

Theorem 7–1–3 [*Barnett and Storey* (*1967*)]

$$|\beta_i| \leqslant \gamma_1, \; i = 1, 2, \ldots, n.$$

Proof. jS is hermitian, since $(jS)^* = -jS' = jS$.

From (7–1–1), $A'P - PA = -2S$

so that
$$jx^*(A'P - PA)x = -2x^*jSx. \tag{7–1–6}$$

Let x be a characteristic vector of A corresponding to a root λ, so that

$$Ax = \lambda x, \; x^*A' = \bar{\lambda}x^*. \tag{7–1–7}$$

Substitution of (7–1–7) into (7–1–6) gives

$$j(\bar{\lambda} - \lambda)x^*Px = -2x^*jSx,$$

whence $2\beta x^*Px = -2x^*jSx$; or $\beta = -x^*jSx/x^*Px$.

Since γ_1 is the greatest root of $jP^{-1}S$ and $-\gamma_1$ the least it follows from Theorem 2–7–5 that $|\beta| \leqslant \gamma_1$.

The magnitudes of the imaginary parts of the characteristic roots of A do not of course affect the stability behaviour of (7–1–4), but Theorem 7–1–3 is nevertheless of some interest.

Thus (7–1–1) or (7–1–3) can be used to construct stability matrices with characteristic roots which are to some extent predetermined. The problem of choosing P, Q, and S so that A has given characteristic roots is related to an important one in control theory [see, e.g., Wonham (1967)] but remains at present unsolved.

Exercise 7–1–1

In (7–1–1) show that $\text{tr}(P^{-1}S) = 0$. Hence deduce that $\sum_{i=1}^{n} \lambda_i(A) = -\sum_{i=1}^{n} \mu_i$; i.e., that the sum of the characteristic roots of A in (7–1–1) is also independent of S.

7–2 Sum of two stability matrices

A sufficient condition is now given for the sum of two stability matrices also to be stable. Let A be a given stability matrix and let P be the corresponding solution of the Liapunov matrix equation (5–3–3) for some given matrix Q. Then:

Theorem 7–2–1 [*Barnett and Storey* (*1967*)]
A sufficient (but not necessary) condition that $A + B$ be stable is that B be of the form

$$B = P^{-1}(S_0 - Q_0), \qquad (7\text{–}2\text{–}1)$$

where S_0 is an arbitrary skew-symmetric matrix and Q_0 an arbitrary symmetric positive semi-definite matrix.

Proof. B is itself a stability matrix, by virtue of (7–1–1), and $A + B$ is a stability matrix by Theorem 5–3–2 since

$$(A + B)'P + P(A + B) = -Q - 2Q_0.$$

If $A + B$ is the matrix of a system of linear differential equations

$$\dot{x} = (A + B)x \qquad (7\text{–}2\text{–}2)$$

then the quadratic form $V = x'Px$ is a Liapunov function for (7–2–2) with $dV/dt = -x'Qx - 2x'Q_0x \leqslant -x'Qx$, so that (7–2–2) is 'at least' as stable as (7–1–4), or 'more' stable if Q_0 is taken positive-definite.

Notice that as in Section 7–1 an equivalent alternative to (7–2–1) is

$$B = (S_0 - Q_0)P, \qquad (7\text{–}2\text{–}3)$$

and by similar reasoning it also follows that the real parts of the characteristic roots of both B and $A + B$ are essentially independent of S_0. Notice also that no matrix $A + B$ of the forms given above can have principal diagonal identical to that of A, unless $Q_0 = 0$ (see Exercise 7–2–1).

Thus the solution of the Liapunov matrix equation (5–3–3) for P provides information not only about the stability behaviour of the system (7–1–4) but also allows the setting up of what may be called 'parametric stability.' That is, given any stability matrix A a family of stability matrices $A + B$ can be constructed with elements in parametric form, the parameters being the elements of S_0, Q_0 and (indirectly) Q. A convenient procedure would be as follows: let ϕ_i be the solution of (5–3–3) when Q has unity as the ith element on the principal diagonal and zero everywhere else. Then if $Q = \text{diag}(\lambda_1, \lambda_2, \ldots, \lambda_n)$ the corresponding solution P of (5–3–3) is $\sum_{i=1}^{n} \lambda_i \phi_i$ with all $\lambda_i \geqslant 0$ and B is then given by (7–2–1) or (7–2–3). This is illustrated by the following simple example [Barnett and Storey (1968*a*)]:

$$A = \begin{bmatrix} 4 & 6 \\ -5 & -7 \end{bmatrix} \text{ is a stability matrix, and taking } Q = \begin{bmatrix} \lambda_1 & 0 \\ 0 & \lambda_2 \end{bmatrix}$$

gives $$P = \frac{1}{12}\begin{bmatrix} 51\lambda_1 + 25\lambda_2 & 42\lambda_1 + 20\lambda_2 \\ 42\lambda_1 + 20\lambda_2 & 36\lambda_1 + 18\lambda_2 \end{bmatrix}.$$

With $S_0 = 12\begin{bmatrix} 0 & \alpha \\ -\alpha & 0 \end{bmatrix}$ and $Q_0 = 12\begin{bmatrix} q_1 & 0 \\ 0 & q_2 \end{bmatrix}$,

Eqn (7–2–3) gives

$$A + B = \begin{bmatrix} 4 & 6 \\ -5 & -7 \end{bmatrix} + \alpha\begin{bmatrix} 42\lambda_1 + 20\lambda_2 & 36\lambda_1 + 18\lambda_2 \\ -51\lambda_1 - 25\lambda_2 & -42\lambda_1 - 20\lambda_2 \end{bmatrix}$$

$$- q_1\begin{bmatrix} 51\lambda_1 + 25\lambda_2 & 42\lambda_1 + 20\lambda_2 \\ 0 & 0 \end{bmatrix} - q_2\begin{bmatrix} 0 & 0 \\ 42\lambda_1 + 20\lambda_2 & 36\lambda_1 + 18\lambda_2 \end{bmatrix}$$

which is a stability matrix for $q_1, q_2, \lambda_1, \lambda_2 \geqslant 0$ and α arbitrary.

Exercise 7–2–1

Show that B in (7–2–3) satisfies

$\text{tr } B = -\text{tr }(Q_0 P) = -\sum_{i=1}^{n} \lambda_i(Q_0 P)$ and hence deduce that the principal diagonal of B cannot be identically zero except when $Q_0 = 0$.

7–3 Jacobi and quasi-Jacobi matrices

Some conditions for stability of Jacobi matrices (Section 3–5) and 'quasi-Jacobi' matrices (to be defined) are now established by considering particular solutions of the Liapunov matrix equation (5–3–3).

Let $T = (t_{ij})$ be an $n \times n$ tridiagonal matrix so that $t_{ij} = 0$ if $|i - j| \geqslant 2$ and

$$d_p(T) = t_{pp}d_{p-1}(T) - t_{p,\,p-1}t_{p-1,\,p}d_{p-2}(T) \tag{7–3–1}$$

where $d_p(T) = \det T(1, 2, \ldots, p \mid 1, 2, \ldots, p)$ (Exercise 3–5–1)

Theorem 7–3–1 [*Barnett and Storey* (*1968b*)]
Sufficient conditions for T to be a stability matrix are $t_{ii} < 0$, $i = 1, 2, \ldots, n$ and $t_{i,\,i+1}t_{i+1,\,i} < 0$, $i = 1, 2, \ldots, n - 1$.

Proof. The matrix

$$P = \operatorname{diag}\left[1, -\frac{t_{12}}{t_{21}}, \frac{t_{12}t_{23}}{t_{21}t_{32}}, \ldots, \frac{(-1)^{n-1}t_{12}t_{23}\cdots t_{n-1,\,n}}{t_{21}t_{32}\cdots t_{n,n-1}}\right]$$

is the solution of the equation $T'P + PT = -Q$ with

$$Q = -2\operatorname{diag}\left[t_{11}, -\frac{t_{12}t_{22}}{t_{21}}, \ldots, \frac{(-1)^{n-1}t_{12}t_{23}\cdots t_{n-1,\,n}t_{nn}}{t_{21}t_{32}\cdots t_{n,n-1}}\right].$$

If the conditions of the theorem are satisfied, P and Q are positive definite, so that by Theorem 5–3–2 T is stable.

Theorem 7–3–2 [*Barnett and Storey* (1968*b*)]
If T satisfies the conditions:

(i) $t_{i,i+1}t_{i+1,\,i} > 0, \quad i = 1, 2, \ldots, n - 1$

(ii) $(-1)^p d_p(T) > 0, \quad p = 1, 2, \ldots, n$

then it is stable.

The conditions (ii) are called by economists the *hicksian* conditions [Metzler (1945)]. Notice that condition (i) implies T is a Jacobi matrix and that all its characteristic roots are real (Section 3–5).

Proof. First, if $E = \operatorname{diag}(e_1, e_2, \ldots, e_n)$ where

$$e_1 = 1,\ e_i = \sqrt{\left(\frac{t_{21}t_{32}\cdots t_{i,\,i-1}}{t_{12}t_{23}\cdots t_{i-1,\,i}}\right)}, \quad i = 2, 3, \ldots, n$$

then the matrix $\tilde{T} = E^{-1}TE$ is symmetric [Genin and Maybee (1966)]. Let

$$R = -E^{-2}T. \tag{7–3–2}$$

This is symmetric since $\tilde{T}$ is symmetric, and

$$T'R + RT = -2T'E^{-2}T. \tag{7–3–3}$$

The matrix on the right of (7–3–3) is positive definite since T is non-singular (Theorem 2–8–7) and $E^{-2} = \operatorname{diag}(f_1, f_2, \ldots, f_n)$

where $f_1 = 1, f_i = \dfrac{t_{12}t_{23}\cdots t_{i-1,i}}{t_{21}t_{32}\cdots t_{i,i-1}}, \quad i = 2, 3, \ldots, n$

and $f_i > 0$ because of (i). The matrix $R = (r_{ij})$ is also Jacobi and

$$r_{ij} = -f_i t_{ij}. \tag{7–3–4}$$

The stability of T is now proved by showing that R is positive definite.

By virtue of (7–3–4), $d_1(R) = -d_1(T)$ and $d_2(R) = f_2 d_2(T)$, and a proof by induction establishes that

$$d_p(R) = (-1)^p f_2 f_3 \cdots f_p d_p(T), \quad p = 2, 3, \ldots, n. \tag{7–3–5}$$

For suppose (7–3–5) holds for $p - 1$. Then from (7–3–1),

$$\begin{aligned} d_p(R) &= r_{pp} d_{p-1}(R) - r_{p,p-1} r_{p-1,p}\, d_{p-2}(R) \\ &= -f_p t_{pp}(-1)^{p-1} f_2 \ldots f_{p-1}\, d_{p-1}(T) \\ &\qquad -f_p f_{p-1} t_{p,p-1} t_{p-1,p}(-1)^{p-2} f_2 \cdots f_{p-2} d_{p-2}(T) \end{aligned}$$

using (7–3–4) and (7–3–5)

$$\begin{aligned} &= (-1)^p f_2 \cdots f_p [t_{pp} d_{p-1}(T) - t_{p,p-1} t_{p-1,p} d_{p-2}(T)] \\ &= (-1)^p f_2 \cdots f_p d_p(T) \quad \text{using (7–3–1).} \end{aligned}$$

Now R is positive definite if and only if $d_1(R), d_2(R), \ldots d_n(R)$ are all positive (Theorem 2–8–5); that is, if and only if T satisfies the conditions (ii), so the theorem is proved.

A crucial property of Jacobi matrices is the recurrence relationship between principal minors (7–3–1). Maybee (1966) has proposed a generalization of Jacobi matrices by defining what he terms *quasi-Jacobi matrices* as follows: An $n \times n$ matrix $A = (a_{ij})$ is quasi-Jacobi if it is indecomposable (Section 2–1) and if for each $p \geqslant 2$,

$$d_p = a_{pp}\, d_{p-1} - \sum_{q=1}^{p-1} a_{pq}\, a_{qp}\, d_{p-2,q},$$

where $d_p = \det A(1, 2, \ldots, p \mid 1, 2, \ldots, p)$ and

$d_{p-2,q} = \det A(1, 2, \ldots, q-1, q+1, \ldots, p-1 \mid 1, 2, \ldots, q-1, q+1, \ldots, p-1)$.

Two properties of quasi-Jacobi matrices among others given by Maybee (1966) are:

Theorem 7–3–3
A quasi-Jacobi matrix is *combinatorially symmetric*; that is, $a_{ij} \neq 0$ implies $a_{ji} \neq 0$, $i, j, = 1, 2, \ldots, n$.

Theorem 7–3–4
A quasi-Jacobi matrix has exactly $n-1$ non-zero elements above the principal diagonal.

Using the Liapunov matrix equation (5–3–3) it is again possible to obtain sufficient conditions for stability analogous to those of Theorem 7–3–1:

Theorem 7–3–5 [*Barnett and Storey* (1968*b*)]
If $H = (h_{ij})$ is quasi-Jacobi there exist diagonal matrices P, Q satisfying

$$H'P + PH = -Q. \tag{7–3–6}$$

Further, if $h_{ij}\,h_{ji} < 0$ when $h_{ij} \neq 0$ and $h_{ii} < 0$ $(i = 1, 2, \ldots, n)$ then H is stable.

Proof. Let $P = \text{diag}\,(p_1, p_2, \ldots, p_n)$. For Q in (7–3–6) to be diagonal, equating the off-diagonal elements to zero gives

$$p_i h_{ij} + p_j h_{ji} = 0 \quad (i \neq j) \tag{7–3–7}$$

for all i, j such that $h_{ij} \neq 0$. Now by Theorem 7–3–4 there are just $(n-1)$ equations in (7–3–7), and by Theorem 2–1–2 since H is indecomposable all the n unknown p_i are present in (7–3–7), so these equations are consistent. Thus a diagonal Q exists and may be denoted by $Q = \text{diag}\,(q_1, q_2, \ldots, q_n)$. From (7–3–6) it follows that $q_i = -2h_{ii}\,p_i$. If p_1 is set equal to unity in (7–3–7) the other p_i $(i = 2, 3, \ldots, n)$ will be positive provided $h_{ij}\,h_{ji} < 0$ and the q_i will then be positive provided $h_{ii} < 0$. Thus both P and Q are positive definite and H is stable by Theorem 5–3–2.

The converse of Theorem 7–3–5 is certainly not true, for take $P_1 = \text{diag}\,(m_1, m_2, \ldots, m_n)$ and $Q = \text{diag}\,(l_1, l_2, \ldots, l_n)$ in (7–1–2) to give a stability matrix $A = (a_{ij})$ with

$$a_{ii} = -m_i l_i, \quad a_{ij} = m_i\, s_{ij} \tag{7–3–8}$$

where $l_i, m_i > 0$ and $s_{ij} = -s_{ji}$. This matrix A satisfies the Liapunov matrix equation with P, Q diagonal but clearly need not be quasi-Jacobi. However, A is combinatorially symmetric and satisfies $a_{ii} < 0$, $a_{ij}a_{ji} \leqslant 0$ so that the stable quasi-Jacobi matrices of Theorem 7–3–5 are a subset of the matrices (7–3–8).

Furthermore, since the matrices P_1 and Q used to construct A in (7–3–8) are diagonal it easily follows from Theorem 7–1–2 that the real parts α_1 of the characteristic roots of A satisfy $\max_k (a_{kk}) \geqslant \alpha_i \geqslant \min_k (a_{kk}), i = 1, 2, \ldots, n.$ That is, the magnitudes of the real parts of the roots of A are essentially determined only by the elements on the principal diagonal of A. In any case provided the off-diagonal elements are as in (7–3–8), A will be a stability matrix for any a_{ii} provided they are all negative. In particular if all the a_{ii} are equal to ($-a$, say) this implies that all the α_i will also be equal to $-a$.

Exercise 7–3–1

The stiffness matrix of a discrete mass-spring system has the form [Genin and Maybee (1966)]

$$M = \begin{bmatrix} m_0 + m_1 & -m_1 & 0 & 0 & \cdots & 0 \\ -m_1 & m_1 + m_2 & -m_2 & 0 & \cdots & 0 \\ 0 & -m_2 & m_2 + m_3 & -m_3 & \cdots & 0 \\ \cdot & \cdot & \cdot & \cdot & \cdots & \cdot \\ \cdot & \cdot & \cdot & \cdot & \cdots & \cdot \\ \cdot & \cdot & \cdot & \cdot & \cdots & \cdot \\ 0 & 0 & 0 & 0 & \cdots & m_{n-1} + m_n \end{bmatrix}$$

with all $m_i > 0$. Use (7–3–1) to prove by induction that

$d_p(M) = [\sum_{j=0}^{p} (1/m_j)] m_0 m_1 m_2 \ldots m_p$, $p = 2, 3, \ldots, n$, and hence deduce that M is positive definite. Take $P_1 = \text{diag}(p_1, p_2, \ldots, p_n)$, $Q = M$ and S to be a Jacobi matrix in (7–1–2) and show that the resulting stability matrix $A = (a_{ij})$ is given by

$$a_{ii} = -(m_{i-1} + m_i)p_i, \qquad i = 1, 2, \ldots, n$$

$$\left.\begin{aligned} a_{i,i+1} &= (m_i + s_{i,i+1})p_i \\ a_{i+1,i} &= (m_i - s_{i,i+1})p_{i+1} \end{aligned}\right\} i = 1, 2, \ldots, n-1$$

$$a_{ij} = 0, \quad j \neq i \pm 1.$$

Notice that $\quad a_{i,i+1}a_{i+1,i} = (m_i^2 - s_{i,i+1}^2)p_i p_{i+1}$

which can be of arbitrary sign.

7–4 Qualitative stability

An important situation in linear economic theory is when no quantitative information is available – that is, only the sign pattern (including zero entries) of matrices involved is known [Quirk and Ruppert (1965)]. Two real matrices $A = (a_{ij})$, $B = (b_{ij})$ are said to be *sign similar* if either sgn $(a_{ij}) = \text{sgn}(b_{ij})$ or $a_{ij} = b_{ij} = 0$, all i, j. Let $\{\bar{A}\}$ denote the set of matrices which are sign similar to a given matrix A. Then A is said to be *sign stable* (or *qualitatively stable*) if and only if every member of $\{\bar{A}\}$ is stable.

Another frequently used notion is that of D-stability: a matrix A is said to be *D-stable* if DA is stable for any choice of matrix $D = \text{diag}(d_{11}, d_{22}, \ldots, d_{nn})$ with all $d_{ii} > 0$. Since pre-multiplying A by the positive diagonal matrix D does not alter the sign pattern of A, it follows at once that:

Theorem 7–4–1

A sign stable matrix is also D-stable.

A useful result is:

Theorem 7–4–2
If there exists a positive diagonal matrix P such that $A'P + PA$ is negative definite, then A is D-stable.

For example, the matrix T of Theorem 7–3–1 is D-stable, as is the matrix H in Theorem 7–3–5 since in both cases the relevant matrix P is positive, diagonal. Indeed, any matrix constructed from (7–1–1) with P diagonal will be D-stable.

Closely related to D-stable matrices are totally stable matrices: a matrix A is said to be *totally-stable* if every principal submatrix $A(1, 2, \ldots, i \mid 1, 2, \ldots, i)$, $i = 1, 2, \ldots, n$ is D-stable. Thus A totally stable $\Rightarrow$ A D-stable $\Rightarrow$ A stable. Metzler (1945) proved the following result:

Theorem 7–4–3
A necessary condition for A to be totally stable is that it satisfies the hicksian conditions

$$(-1)^p d_p(A) > 0, \quad p = 1, 2, \ldots, n.$$

Notice that this is a stronger version of Theorem 7–3–2.

Also useful is:

Theorem 7–4–4
If $a_{ii} < 0, \quad i = 1, 2, \ldots, n$, then A sign stable $\Rightarrow$ A totally stable.

All the above theorems may be found in the paper by Quirk and Ruppert (1965) together with a full discussion and a number of other results, of which the most general is:

Theorem 7–4–5
Let $A = (a_{ij})$ be an $n \times n$ indecomposable matrix. Then necessary and sufficient conditions for A to be sign stable are:

(i) $a_{ij}a_{ji} \leqslant 0, \quad i \neq j$
(ii) $i_1 \neq i_2 \neq \cdots \neq i_m$, $a_{i_1 i_2} \neq 0$, $a_{i_2 i_3} \neq 0, \ldots, a_{i_{m-1}, i_m} \neq 0$ implies $a_{i_m i_1} = 0$ for any $m > 2$.
(iii) $a_{ii} \leqslant 0$ for all i, $a_{kk} < 0$ for at least one k.
(iv) there exists a non-zero term in the expansion of det A.

Other related results and definitions may be found in papers by Carlson (1968), Maybee and Quirk (1969), Maybee (1966), and Ostrowski and Schneider (1962), and a brief review of work in the field is given by Taussky (1964).

Exercise 7–4–1 [Quirk and Ruppert (1965)]
If A is sign stable show that sgn det $A = (-1)^n$. Hence, show that if A_{ij} is the cofactor of a_{ij} then sgn $A_{ij} = (-1)^n$ sgn a_{ij} provided $a_{ij} \neq 0$, $A_{ij} \neq 0$.

Exercise 7–4–2
Let $X = S - kI$ where S is skew-symmetric and k a real positive scalar, and let $A = XD$ where D is a diagonal matrix. Use the Liapunov matrix equation to show directly that if all $d_{ii} > 0$, A is stable [the converse also holds, see Taussky (1961)].

Discussion and further reading

Liapunov's basic theorem for linear systems has been used to derive results on the theory of stability matrices, thus extending the scope of the second method. The novel approach to algebraic problems so provided is particularly useful since fairly powerful theorems may be obtained without the necessity for complex proofs. For example, it would be very difficult to obtain by conventional matrix methods even the sufficient conditions for construction of a stable matrix (Theorem 7–1–1). The work on parametric stability is also valuable since in general little can be said about the variation in the characteristic roots of a matrix if the elements change by finite amounts [see Taussky (1964)]. These remarks are further illustrated by the applications to Jacobi and quasi-Jacobi matrices.

References

BARNETT, S. and C. STOREY (1967), Analysis and synthesis of stability matrices, *J. Diff. Eqns*, **3,** 414–422.

BARNETT, S. and C. STOREY (1968*a*), Some results on the sensitivity and synthesis of asymptotically stable linear and non-linear systems, *Automatica*, **4,** 187–194.

BARNETT, S. and C. STOREY (1968*b*), Some applications of the Liapunov matrix equation, *J. Inst. Math. Appl.*, **4,** 33–42.

CARLSON, D. (1968), A new criterion for H-stability of complex matrices, *Linear Algebra*, **1,** 59–64.

GENIN, J. and J. S. MAYBEE (1966), A stability theorem for a class of damped dynamic systems and some applications, *J. Inst. Math. Appl.*, **2,** 343–357.

MAYBEE, J. S. (1966), New generalizations of Jacobi matrices, *SIAM J. Appl. Math.*, **14,** 1032–1037.

MAYBEE, J. S. and J. P. QUIRK (1969), Qualitative problems in matrix theory, *SIAM Rev.*, **11,** 30–51.

METZLER, L. (1945), Stability of multiple markets: the Hicks conditions, *Econometrica*, **13,** 277–292.

OSTROWSKI, A. and H. SCHNEIDER (1962), Some theorems on the inertia of general matrices, *J. Math. Anal. Appl.*, **4,** 72–84.

QUIRK, J. P. and R. RUPPERT (1965), Qualitative economics and the stability of equilibrium, *Rev. Econ. Studies*, **32,** 311–326.

TAUSSKY, O. (1961), A remark on a theorem of Liapunov, *J. Math. Anal. Appl.*, **2,** 105–107.

TAUSSKY, O. (1964), On the variation of the characteristic roots of a finite matrix under various changes of its elements, *Recent Advances in Matrix Theory*, ed. H. Schneider, University of Wisconsin Press, 125–138.

VOGT, W. G. (1965), Transient response from the Liapunov stability equation, *Preprints Joint Automatic Control Conf.*, Paper V 4, 23–30.

WONHAM, W. M. (1967), On pole assignment in multi-input controllable linear systems, *I.E.E.E. Trans. Aut. Control.*, **AC–12**, 660–665.

8

Extensions to non-linear systems

8–1 The method of linear bounds

A basic theorem of Liapunov extends the theory of stability of linear systems

$$\dot{x} = Ax, \tag{8–1–1}$$

where x is a column n-vector and A a real constant matrix, to that of systems of the form

$$\dot{x} = Ax + g(x), \quad g(0) = 0, \tag{8–1–2}$$

where $g(x) = [g_1(x), g_2(x), \ldots, g_n(x)]'$.

Suppose the functions $g_i(x)$ have power series expansions in the neighbourhood of the origin in the variables $x_1, x_2, \ldots, x_n$ beginning with terms of at least second degree. Then (8–1–1) is called the *first approximation* (or *linear part*) of (8–1–2) and is said to be obtained from (8–1–2) by *linearization.* In this case:

Theorem 8–1–1 (Liapunov)
If (8–1–1) is asymptotically stable at the origin so is (8–1–2). If (8–1–1) is unstable (i.e., A has at least one characteristic root with positive real part) then (8–1–2) is also unstable.

A proof of Theorem 8–1–1, together with related theorems and examples is given by Lehnigk (1966). In the critical case the stability behaviour is not determined by first-degree terms in (8–1–2) and further investigation is required [Liapunov (1966)].

Hahn (1963) also gives Theorem (8–1–1) but points out that the assumption made about $g(x)$ is very restrictive in practice; he provides an improvement, the so-called 'method of linear bounds,' and this is now described.

Consider first the auxiliary linear system

$$\dot{x} = (A + G)x \tag{8–1–3}$$

where G is a constant matrix to be determined. Let $V = x'Px$ be a Liapunov function for (8–1–1) with $A'P + PA = -Q$, P and Q being positive definite. The derivative of V with respect to (8–1–3) is $-x'Rx$ where

$$R = Q - G'P - PG. \tag{8–1–4}$$

If (8–1–3) is to be asymptotically stable with Liapunov function V then R must be positive definite, and the conditions (see Theorem 2–8–5) for this to be so provide n inequalities which must be satisfied by the n^2 elements g_{ij} of G. Since $G \equiv 0$ makes (8–1–3) asymptotically stable there will be a solution to these inequalitities, giving (say) $g_{ij}^{(1)}$ and $g_{ij}^{(2)}$ as the end points of the interval within which g_{ij} must lie ($i, j = 1, 2, \ldots, n$).

We now return to the non-linear system (8–1–2). It follows that V will also be a Liapunov function for (8–1–2) provided each $g_i(x)$ satisfies linear inequalities

$$x'g_i^{(1)} \leqslant g_i(x) \leqslant x'g_i^{(2)} \text{ or } x'g_i^{(1)} \geqslant g_i(x) \geqslant x'g_i^{(2)}, \; i = 1, 2, \ldots, n \tag{8–1–5}$$

where $g_i^{(1)}$ and $g_i^{(2)}$ are column n-vectors with components $g_{ij}^{(1)}$ and $g_{ij}^{(2)}$ respectively. Thus:

Theorem 8–1–2 [*Hahn* (*1963*)].
The origin of system (8–1–2) is asymptotically stable with Liapunov function V provided that each $g_i(x)$ satisfies the *linear bounds* (8–1–5).

Nesbit (1965) has pointed out that if more than one component of g is non-zero then Theorem 8–1–2 must be modified. However, in practice many systems are in phase-variable form (5–1–4) in which case the difficulty does not arise.

It is in fact possible to give an explicit expression for functions $g(x)$ which ensure that the origin of the system (8–1–2) is asymptotically stable. First, we write (8–1–4) in the form

$$G = P^{-1}(\tfrac{1}{2}Q - \tfrac{1}{2}R + S), \tag{8–1–6}$$

where S is an arbitrary skew-symmetric matrix. The system (8–1–3) will then be asymptotically stable with G given by (8–1–6) for any choice of positive definite R. Next, notice that the elements of R and S in (8–1–6) can be functions of x provided that R remains symmetric positive definite,* S remains skew-symmetric and $Gx = 0$ when $x = 0$. Thus:

Theorem 8–1–3 [*Barnett and Storey* (*1966a*)]
The origin of system (8–1–2) is asymptotically stable with Liapunov function $V = x'Px$ having $\dot{V} = -x'Rx$ for all functions of the form $g(x) = P^{-1}(\tfrac{1}{2}Q - \tfrac{1}{2}R + S)x$ where P, Q are constant positive definite matrices satisfying $A'P + PA = -Q$, $R(x)$ is an arbitrary symmetric positive definite matrix and $S(x)$ an arbitrary skew-symmetric matrix.

An example used by Nesbit (1965) is now given to illustrate the method. The equation

$$\ddot{x} + \dot{x} + x - g(x, \dot{x}) = 0 \tag{8–1–7}$$

* A matrix $R(x, t)$ is called positive definite if the function $x'Rx$ is positive definite (see Definition 5–2–2).

where x is a scalar variable, can be written

$$\begin{bmatrix}\dot{x}_1\\ \dot{x}_2\end{bmatrix}=\begin{bmatrix}0 & 1\\ -1 & -1\end{bmatrix}\begin{bmatrix}x_1\\ x_2\end{bmatrix}+\begin{bmatrix}0\\ g\end{bmatrix},$$

by taking $x = x_1$, $\dot{x} = x_2$. In order that a comparison may be made with Nesbit's expression for g, take Q in Theorem 8–1–3 to be the unit matrix and add the further requirement $\dot{V} < -\frac{1}{2}x'x$. Setting $R - \frac{1}{2}I = T$ gives $\dot{V} = -\frac{1}{2}x'x - x'Tx$ so T is to be positive definite. From (8–1–6)

$$G = P^{-1}(\tfrac{1}{4}I - \tfrac{1}{2}T + S). \tag{8–1–8}$$

The solution of $A'P + PA = -I$ is

$$P=\begin{bmatrix}\frac{3}{2} & \frac{1}{2}\\ \frac{1}{2} & 1\end{bmatrix} \quad\text{and}\quad P^{-1}=\tfrac{4}{5}\begin{bmatrix}1 & -\frac{1}{2}\\ -\frac{1}{2} & \frac{3}{2}\end{bmatrix}.$$

Let
$$S=\begin{bmatrix}0 & \alpha\\ -\alpha & 0\end{bmatrix},\quad T=\begin{bmatrix}t_1 & t_2\\ t_2 & t_3\end{bmatrix},\quad G=\begin{bmatrix}0 & 0\\ g_{21} & g_{22}\end{bmatrix}.$$

Eqn (8–1–8) then gives

$$\left.\begin{aligned}0 &= \tfrac{1}{4} - \tfrac{1}{2}t_1 + \tfrac{1}{4}t_2 + \tfrac{1}{2}\alpha\\ 0 &= -\tfrac{1}{8} - \tfrac{1}{2}t_2 + \tfrac{1}{4}t_3 + \alpha\end{aligned}\right\} \tag{8–1–9}$$

$$g_{21} = \tfrac{4}{5}(-\tfrac{1}{8} + \tfrac{1}{4}t_1 - \tfrac{3}{4}t_2 - \tfrac{3}{2}\alpha) \tag{8–1–10}$$

$$g_{22} = \tfrac{4}{5}(\tfrac{3}{8} + \tfrac{1}{4}t_2 - \tfrac{3}{4}t_3 - \tfrac{1}{2}\alpha). \tag{8–1–11}$$

Solving (8–1–9) for t_1 and t_3 in terms of t_2 and α and substituting into (8–1–10) and (8–1–11) gives

$$\left.\begin{aligned}g_{21} &= -(\alpha + \tfrac{1}{2}t_2)\\ g_{22} &= 2\alpha - t_2\end{aligned}\right\}. \tag{8–1–12}$$

One of the conditions for positive definiteness of T, namely $t_1t_3 > t_2^2$ leads to $t_2 > \frac{4}{5}(4\alpha - \frac{1}{2})(\alpha + \frac{1}{2})$. Thus, setting $\beta = t_2 - \frac{4}{5}(4\alpha - \frac{1}{2})(\alpha + \frac{1}{2})$ enables (8–1–12) to be written

$$\left.\begin{aligned}g_{21} &= -\tfrac{8}{5}\alpha^2 - \tfrac{8}{5}\alpha + \tfrac{1}{10} - \tfrac{1}{2}\beta\\ g_{22} &= -\tfrac{16}{5}\alpha^2 + \tfrac{4}{5}\alpha + \tfrac{1}{5} - \beta\end{aligned}\right\} \tag{8–1–13}$$

where α is arbitrary and $\beta > 0$. The condition $t_1 > 0$ does not lead to any further restrictions on α or β, so that by Theorem 8–1–3, the origin of (8–1–7) is asymptotically stable for functions

$$g(x, \dot{x}) = (\tfrac{1}{10} - \tfrac{8}{5}\alpha^2 - \tfrac{8}{5}\alpha - \tfrac{1}{2}\beta)x + (\tfrac{1}{5} - \tfrac{16}{5}\alpha^2 + \tfrac{4}{5}\alpha - \beta)\dot{x}$$

subject only to $\beta(x, \dot{x}) > 0$ and $g(0, 0) = 0$.

By finding the maximum of each of the expressions in (8–1–13) with respect to α by simple differentiation it is easily found that

$$g_{21} < \tfrac{1}{2},\qquad g_{22} < \tfrac{1}{4}. \tag{8–1–14}$$

Also, elimination of α and β between the two expressions in (8–1–13) gives

$$(g_{21} - \tfrac{1}{2}g_{22})^2 + (g_{22} + \tfrac{1}{2}g_{21}) < \tfrac{1}{4}. \tag{8–1–15}$$

It is interesting that the inequalities (8–1–14) and (8–1–15) satisfied by the g_{ij} are identical with those obtained by Nesbit using Theorem 8–1–2, but the greater generality of Theorem 8–1–3 is evident from the fact that α and β in (8–1–13) can be functions of x and $\dot{x}$.

8–2 General non-linear systems: sensitivity and synthesis

The power and usefulness of the method of the previous section are now demonstrated by the ease with which it is possible to generalize to non-linear systems.

Suppose that the origin of the non-linear system

$$\dot{x} = f(x), \qquad f(0) = 0 \tag{8–2–1}$$

is asymptotically stable with a known Liapunov function $V(x)$ having $dV/dt = -N < 0$.

Then the basic sensitivity theorem is:

Theorem 8–2–1 [*Barnett and Storey* (*1966b*)]
The origin of the system

$$\dot{x} = f(x) + h(x) \tag{8–2–2}$$

will also be asymptotically stable for all functions $h(x)$ of the form

$$h(x) = (S - Q)\phi, \tag{8–2–3}$$

where $S \equiv S(x)$ is an arbitrary skew-symmetric matrix, $Q \equiv Q(x)$ is an arbitrary positive semi-definite matrix, $\phi(x) = \nabla V$ and $h(0) = 0$.

Proof. $V(x)$ is a Liapunov function for (8–2–3), since

$$dV/dt = -N - \phi' Q \phi \leqslant -N.$$

In the linear case $f(x) = Ax$, $V = x'Px$, and $\phi = 2Px$ so (8–2–3) reduces to (7–1–3). As in Section 7–1, it is of interest to point out that the choice (8–2–3) for h makes (8–2–2) at least as stable as (8–2–1). However, it should be remembered that (8–2–3) provides only a sufficient and not necessary condition for the origin of (8–2–2) to be stable.

A similar result holds for the non-autonomous system

$$\dot{x} = f(x, t), \qquad f(0, t) = 0$$

with a known decrescent (Definition 5–2–7) Liapunov function $V(x, t)$, except that S and Q in (8–2–3) may also be dependent upon t.

An illustrative example [Schultz (1965)] is now given. The system described by

$$\ddot{x} + \dot{x} + x^3 = 0.$$

or

$$\dot{x}_1 = x_2, \quad \dot{x}_2 = -x_1^3 - x_2$$

is asymptotically stable at the origin with Liapunov function

$$V = \tfrac{1}{2}x_1^4 + x_1^2 + 2x_1x_2 + x_2^2, \qquad \dot{V} = -2x_1^4.$$

Barbashin's theorem (Theorem 5–2–5) applies since $\dot{V}$ is not zero along a trajectory. Choosing $S = 0$,

$$Q = \tfrac{1}{2}\begin{bmatrix} 0 & 0 \\ 0 & q \end{bmatrix}$$

in (8–2–3) gives

$$h(x) = \begin{bmatrix} 0 \\ -q(x_1 + x_2) \end{bmatrix}$$

which produces, on substitution,

$$\ddot{x} + \dot{x}(1 + q) + x^3 + qx = 0$$

which is asymptotically stable at the origin for any $q(x) \geqslant 0$. V is still a Liapunov function and for the new system

$$\dot{V} = -2x_1^4 - 2q(x_1 + x_2)^2.$$

The result on synthesis of asymptotically stable non-linear systems can now be stated:

Theorem 8–2–2 [*Barnett and Storey* (*1966b*)]
The origin of the system

$$\dot{x} = [S(x, t) - Q(x, t)]\phi(x), \tag{8–2–4}$$

where $S(x, t)$ is skew-symmetric for all t, $Q(x, t)$ is positive definite for all t and $\phi(x)$ is the gradient of any positive definite function $V(x)$, is asymptotically stable.

Proof. The proof is immediate, since V is a Liapunov function with $\dot{V} = -\phi' Q\phi$.

If V is taken to be $x'Px$ (8–2–4) reduces to the linear case (7–1–3). It should be clear also how the results of this section can be modified for stability instead of asymptotic stability.

Some similar but less general results than those of this section have been obtained by Liu and Leake (1967), who keep both the Liapunov function and its derivative fixed.

Exercise 8–2–1

Use the Liapunov function $V = x_1^2 + x_2^2$ to show that the system $\dot{x}_1 = -x_1 + 2x_2$, $\dot{x}_2 = -2x_1 - x_2$ is asymptotically stable. Apply Theorem 8–2–1 to show that the second equation can be $\dot{x}_2 = -2x_1 - x_2 - qx_2$ for any $q(x_1, x_2) \geqslant 0$ without affecting the system stability.

Illustrate the non-necessity of Theorem 8–2–1 by showing that if $q = -x_2$, the system is asymptotically stable in the region $V \leqslant 1$. [Schultz and Melsa (1967), p. 190.]

Exercise 8–2–2

Taking $V(x) = \frac{1}{2}(x_1^2 + x_2^2)$ in Theorem 8–2–2 choose suitable S and Q to show that the origin of the system $\dot{x}_1 = -x_1^3 - x_2$, $\dot{x}_2 = x_1 - x_2^3$ is asymptotically stable.

8–3 Control systems

First consider the linear control system

$$\dot{x} = Ax + Bu, \tag{8–3–1}$$

where A is $n \times n$, B is $n \times m$, and u is the vector of control variables. Assume that A is a stability matrix and let P be the solution of $A'P + PA = -Q$. Then:

Theorem 8–3–1 [*Barnett and Storey* (*1966b*)]

If the control

$$u(x, t) = [S(x, t) - K(x, t)]B'Px, \tag{8–3–2}$$

where $S(x, t)$ is an arbitrary $m \times m$ skew-symmetric matrix and $K(x, t)$ an arbitrary $m \times m$ symmetric positive definite matrix, is applied to (8–3–1) the origin of the resulting system is always asymptotically stable for all choices of S and K.

Proof. This follows familiar lines: the control (8–3–2) gives

$$\dot{x} = Ax + B(S - K)B'Px \tag{8–3–3}$$

and $V = x'Px$ is a Liapunov function for (8–3–3) with

$$\dot{V} = -x'Qx - 2x'PBKB'Px \leqslant -x'Qx.$$

Thus the system with control is more stable than the uncontrolled system $u = 0$.

Setting $S = 0$ and $K = K(t)$ in (8–3–2) gives a result similar to one obtained by Johnson (1964).

An application of this technique to a situation discussed by Grayson (1965) is of interest. He describes a design procedure for systems of the form

$$\dot{x} = f(x) + u, \tag{8–3–4}$$

where the uncontrolled system is stable but not asymptotically stable. This implies under certain conditions on f (see Theorem 5–2–8) that a Liapunov function $U(x)$ exists for the unforced system. It follows that taking

$$u(x) = [S(x) - Q(x)]\psi(x), \tag{8–3–5}$$

where $\psi(x) = \text{grad } U$, makes (8–3–4) asymptotically stable. Grayson considers as an example a problem where the equations of motion are

$$\left.\begin{aligned} \dot{\omega}_1 &= \frac{I_2 - I_3}{I_1}\omega_2\omega_3 + u_1 \\ \dot{\omega}_2 &= \frac{I_3 - I_1}{I_2}\omega_1\omega_3 + u_2 \\ \dot{\omega}_3 &= \frac{I_1 - I_2}{I_3}\omega_1\omega_2 + u_3 \end{aligned}\right\} \tag{8–3–6}$$

where I_i, ω_i, and u_i ($i = 1, 2, 3$) are the moment of inertia, angular rate and control torque respectively about the ith principal axis. A Liapunov function for (8–3–6) is the total rotational energy $U = \frac{1}{2}(I_1\omega_1^2 + I_2\omega_2^2 + I_3\omega_3^2)$ for which $\dot{U} = 0$. Use of (8–3–5) gives the control

$$(u_1, u_2, u_3)' = [S(x) - Q(x)](I_1\omega_1, I_2\omega_2, I_3\omega_3)' \tag{8–3–7}$$

where $S(x)$ and $Q(x)$ are arbitrary. The solution given by Grayson is in fact a special case of (8–3–7) with $S = 0$ and $Q = \text{diag}(\alpha/2I_1, \alpha/2I_2, \alpha/2I_3)$, α being a positive constant.

Discussion and further reading

The ideas and methods of the previous chapter have been extended to deal with non-linear systems, thus highlighting two important points. First, since Liapunov's method applies equally well to both linear and non-linear systems, generalization of the results of Chapter 7 is easy and natural whereas conventional linear system techniques are of little, if any, value in dealing with the non-linear case. Secondly, for a given non-linear system the difficulty will lie in finding a suitable Liapunov function. Although various methods exist for obtaining such functions, the region of stability in phase space thereby obtained may be inadequately small. This chapter has shown how a known Liapunov function can be used to describe the sensitivity of the stability behaviour of non-linear systems to finite perturbations so that effort expended in finding a Liapunov function becomes more worth while [see also Barnett and Storey (1968) and for a survey with many references, Barnett (1968)].

References

Barnett, S. (1968), Algebraic methods in Liapunov theory, *Control*, **12,** 545–549.

Barnett, S. and C. Storey (1966*a*), Asymptotic stability and the method of linear bounds, *Electronics Letters*, **2,** 343.

BARNETT, S. and C. STOREY (1966*b*), Asymptotic stability of non-linear systems, *Electronics Letters*, **2,** 375.

BARNETT, S. and C. STOREY (1968) Some results on the sensitivity and synthesis of asymptotically stable linear and non-linear systems, *Automatica*, **4,** 187–194.

GRAYSON, L. P. (1965), The status of synthesis using Liapunov's method, *Automatica*, **3,** 91–121.

HAHN, W. (1963), *Theory and application of Liapunov's direct method*, Prentice-Hall, Englewood Cliffs, N.J.

JOHNSON, G. W. (1964), Synthesis of control systems with stability constraints via the direct method of Liapunov, *Preprints Joint Automatic Control Conf.*, 269–277.

LEHNIGK, S. H. (1966), *Stability theorems for linear motions with an introduction to Liapunov's direct method*, Prentice-Hall, Englewood Cliffs, N.J.

LIAPUNOV, A. M. (1966), *Stability of motion*, Academic Press, New York.

LIU, R. and R. J. LEAKE (1967) Inverse Liapunov problems, *Differential Equations and Dynamic Systems*, ed. J. K. Hale and J. P. La Salle, Academic Press, New York 333–339.

NESBIT, R. A. (1965), Several applications of the direct method of Liapunov, *Advances in Control Systems*, Vol. 2, Academic Press, New York, 269–308.

SCHULTZ, D. G. (1965), The generation of Liapunov functions, *Advances in Control Systems*, Vol. 2, Academic Press, New York, 1–64.

SCHULTZ, D. G. and J. L. MELSA, (1967), *State functions and linear control systems*, McGraw-Hill, New York.

9

Applications to optimal control theory

9–1 Optimal linear control systems with constant coefficients

Consider the system

$$\dot{x} = Ax + Bu, \tag{9–1–1}$$

where A is a constant $n \times n$ matrix, B a constant $n \times m$ matrix ($m \leqslant n$) and the vector u, known as the control vector [Noton (1965)] is to be chosen so as to minimize a *quadratic performance index*

$$I = \int_0^T (x'Qx + u'Ru)\, dt, \tag{9–1–2}$$

the matrices Q and R being constant symmetric positive definite and of appropriate dimensions. The final time T and the initial state $x(0)$ are assumed fixed. Then it is well known [Kalman (1960)] that the optimal control vector is given by

$$u = -R^{-1}B'Px, \tag{9–1–3}$$

where $P(t)$ is the symmetric positive definite solution of the matrix Riccati equation

$$\dot{P} = PBR^{-1}B'P - A'P - PA - Q, \tag{9–1–4}$$

subject to the condition $P(T) = 0$, and the minimum value of I is then

$$J = x'(0)P(0)x(0). \tag{9–1–5}$$

Eqn (9–1–4) can be solved by standard techniques for numerical solution of sets of differential equations – see, e.g., Merriam (1964). (Under certain conditions the solution $P(t)$ is unique, Kalman, 1964.)

For the rest of this chapter it will be assumed that $T = \infty$, in which case (9–1–4) reduces to the so-called 'steady-state' Riccati (algebraic) equation

$$PBR^{-1}B'P - A'P - PA - Q = 0, \tag{9–1–6}$$

the solution P then of course being a constant matrix. Eqn (9–1–6) represents $\frac{1}{2}n(n + 1)$ simultaneous quadratic equations and a computer program [Barnett, Ford, and Storey (1968)] based on a method due to Potter (1966) has been successfully used for problems with $n = 4$ and 5. In fact, because of its importance in control theory, considerable attention has been paid to the problem of solving Eqn (9–1–4) including the more general case when A, B, Q

and R, are all time-varying, and papers by Friedland (1967), Porter (1967), and Blackburn and Bidwell (1968) should be consulted for further details and other references.

Exercise 9–1–1 [Reid (1959)]
If $X(t), Y(t)$ is a solution of the system $\dot{X} = AX - BR^{-1}B'Y, \dot{Y} = -QX - A'Y$, show that $P = YX^{-1}$ is a solution of (9–1–4).

9–2 Sensitivity of optimal linear control systems to small variations in parameters

If $\theta = (\theta_1, \theta_2, \ldots, \theta_k)$ represents some set of physical parameters affecting the matrices in (9–1–1) it is the aim of sensitivity analysis, broadly speaking, to describe in some way the effect on the solution when these parameters alter. For an introduction to sensitivity analysis of dynamic systems see the book by Tomovic (1963); a wide-ranging review of sensitivity in automatic control theory is given by Kokotovic and Rutman (1965); and a review of sensitivity in optimal control is contained in survey papers by Athans (1966) and Sobral (1968).

Following Pagurek (1965), the change in J in (9–1–5) corresponding to small changes $\delta\theta_i$ in the parameters from their original values is, to first order, $\delta J = \sum_i (\partial J/\partial\theta_i)\delta\theta_i$, the partial derivatives being evaluated with θ at its nominal value. By virtue of (9–1–5) it follows that

$$\frac{\partial J}{\partial \theta_i} = \left(x' \frac{\partial P}{\partial \theta_i} x\right)_{t=0} \tag{9–2–1}$$

so the sensitivity of the minimum value of the performance index can be determined by finding the matrices of partial derivatives $\partial P/\partial\theta_i$.

We assume throughout that no changes occur in the matrices Q and R. Then differentiation of (9–1–6) with respect to θ_i gives

$$\frac{\partial P}{\partial \theta_i} BR^{-1}B'P + PBR^{-1}B' \frac{\partial P}{\partial \theta_i} + P\left[\frac{\partial}{\partial \theta_i}(BR^{-1}B')\right]P \\ - A'\frac{\partial P}{\partial \theta_i} - \frac{\partial P}{\partial \theta_i}A - \frac{\partial A'}{\partial \theta_i}P - P\frac{\partial A}{\partial \theta_i} = 0. \tag{9–2–2}$$

Eqn (9–2–2) can then, at least in theory, be solved for $\partial P/\partial\theta_i$. However, for this approach to sensitivity analysis to be of practical utility then clearly the work involved must be less than that needed to recalculate the solution to the changed situation.

Bearing this last point in mind, considerable simplification can be achieved if it is assumed [Barnett (1966)] that the parameter θ_i can be divided into two sets, labelled for convenience $a = \{a_i\}$ and $b = \{b_i\}$, in such a way that A is independent of any of the b_i and B is independent of any of the a_i; that is $\partial A/\partial b_i = 0$ and $\partial B/\partial a_i = 0$ for all possible values of i. In this case Eqn

(9–2–2) is greatly simplified so that solving for $\partial P/\partial\theta_i$ becomes a more practical and worth-while proposition. This separation of parameters into two sets cannot always be done, of course, but an obvious and very important example is when the elements a_{ij} and b_{ij} of A and B are themselves regarded as the parameters. In this latter case Eqn (9–2–1) becomes simplified still further. In general, however, for the assumption concerning the parameters to hold it is only necessary that the (small) changes in A and B occur independently. This is certainly not an unreasonable requirement.

Let A_0, B_0, P_0 stand for the respective matrices when the parameter vectors are at their original values a_0 and b_0, and let the corresponding matrix* in (9–1–3) be $D_0 = -R^{-1}B_0'P_0$. Small changes δa_i and δb_j in the parameters give rise to a corresponding change, to first order, in P of

$$\delta P = \sum_i \left(\frac{\partial P}{\partial a_i}\right)_0 \delta a_i + \sum_j \left(\frac{\partial P}{\partial b_j}\right)_0 \delta b_j$$

and to a change in J

$$\delta J = \sum_i \left(\frac{\partial J}{\partial a_i}\right)_0 \delta a_i + \sum_j \left(\frac{\partial J}{\partial b_j}\right)_0 \delta b_j \qquad (9\text{–}2\text{–}3)$$

the suffix $(\,)_0$ indicating that the derivatives are evaluated with the parameters at their original values. The derivatives of J in (9–2–3) are given by (9–2–1).

Differentiating the steady-state matrix Riccati equation (9–1–6) with respect to a_i and setting $a = a_0$, $b = b_0$, we obtain

$$\left(\frac{\partial P}{\partial a_i}\right)_0 (A_0 + B_0D_0) + (A_0 + B_0D_0)'\left(\frac{\partial P}{\partial a_i}\right)_0 = -\left(\frac{\partial A}{\partial a_i}\right)_0' P_0 - P_0\left(\frac{\partial A}{\partial a_i}\right)_0 \qquad (9\text{–}2\text{–}4)$$

and similarly with respect to b_j,

$$\left(\frac{\partial P}{\partial b_j}\right)_0 (A_0 + B_0D_0) + (A_0 + B_0D_0)'\left(\frac{\partial P}{\partial b_j}\right)_0 = P_0\left(\frac{\partial B}{\partial b_j}\right)_0 R^{-1}B_0'P_0 + P_0B_0R^{-1}\left(\frac{\partial B}{\partial b_j}\right)_0' P_0. \qquad (9\text{–}2\text{–}5)$$

Since the original optimal system (9–1–1) with constant linear control $u = D_0x$ is assumed asymptotically stable, the matrix $A_0 + B_0D_0$ is a stability matrix. Thus Eqns (9–2–4) and (9–2–5) are in exactly the form of the Liapunov matrix equation and can be solved for $(\partial P/\partial a_i)_0$ and $(\partial P/\partial b_j)_0$ using an appropriate method from Chapter 5.

In the special cases when a_i is an element of A, or b_j an element of B, $\partial A/\partial a_i$ or $\partial B/\partial b_j$ have just a single unit element, all other elements being zero, so that the right sides of (9–2–4) and (9–2–5) become extremely simple.

* Known as the 'feedback matrix'.

It is worth noting that zero elements in $\partial P/\partial a_i$ or $\partial P/\partial b_j$ mean that the corresponding elements (i.e., in the same positions) of P are insensitive to small changes in a_i or b_j respectively.

An expression for $\partial D/\partial \theta_i$, the matrix representing sensitivity of the optimal control law is, to first order,

$$\delta D = \sum_i \left(\frac{\partial D}{\partial a_i}\right)_0 \delta a_i + \sum_j \left(\frac{\partial D}{\partial b_j}\right)_0 \delta b_j,$$

where $(\partial D/\partial a_i)_0$ and $(\partial D/\partial b_j)_0$ are given by differentiating the expression $D = -R^{-1}B'P$. The change in the optimal control vector is just $\delta u = (\delta D)x$.

Finally, we derive expressions for $\partial x/\partial \theta_i$, the sensitivity coefficients of the state vector $x(t)$ as follows.

The original optimal system is

$$\dot{x} = (A_0 + B_0 D_0)x. \tag{9–2–6}$$

For small changes in a and b the state vector is, to first order,

$$\begin{aligned} x(t) &= x(a_0 + \delta a, b_0 + \delta b, t) \\ &= x_0(t) + \sum_i \left(\frac{\partial x}{\partial a_i}\right)_0 \delta a_i + \sum_j \left(\frac{\partial x}{\partial b_j}\right)_0 \delta b_j, \end{aligned}$$

where $x_0(t)$ is the solution of (9–2–6). Differentiation of (9–2–6) with respect to a_i gives

$$\frac{d}{dt}\left(\frac{\partial x}{\partial a_i}\right) = (A_0 + B_0 D_0)\left(\frac{\partial x}{\partial a_i}\right) + \left[\left(\frac{\partial A}{\partial a_i}\right)_0 + B_0\left(\frac{\partial D}{\partial a_i}\right)_0\right]x_0. \tag{9–2–7}$$

Simultaneous solution of (9–2–6) and (9–2–7) then gives $(\partial x/\partial a_i)_0$, and $(\partial x/\partial b_j)_0$ can be dealt with in a similar way. It is assumed that the initial conditions for (9–1–1) are unaffected by parameter variations, so that $\partial x/\partial a_i = 0$ at $t = 0$.

Exercise 9–2–1 [Barnett (1966)]

Let $\bar{x}(s)$, $\bar{X}(s)$ denote the Laplace transforms of $x(t)$, $\partial x/\partial a_i$ respectively. Transform Eqns (9–2–6) and (9–2–7) to obtain

$$\bar{X} = (sI - A_0 - B_0 D_0)^{-1}\left[\left(\frac{\partial A}{\partial a_i}\right)_0 + B_0\left(\frac{\partial D}{\partial a_i}\right)_0\right](sI - A_0 - B_0 D_0)^{-1}x(0).$$

Exercise 9–2–2 [Barnett (1966)]

Let $\quad A_0 = \begin{bmatrix} -1 & 0 \\ 1 & 0 \end{bmatrix}, \quad B_0 = \begin{bmatrix} 1 \\ 0 \end{bmatrix}, \quad Q_0 = \begin{bmatrix} 0 & 0 \\ 0 & 1 \end{bmatrix}, \quad R_0 = \frac{1}{10}.$

Show that the corresponding solution of (9–1–6) is

$P_0 = \begin{bmatrix} 0{\cdot}17 & 0{\cdot}316 \\ 0{\cdot}316 & 0{\cdot}85 \end{bmatrix}$. Write $B = \begin{bmatrix} b_1 \\ 0 \end{bmatrix}$ and investigate sensitivity in the neighbourhood of $b_1 = 1$ to show that

$$\left[\frac{\partial P}{\partial b_1}\right]_0 = -\begin{bmatrix} 0{\cdot}22 & 0{\cdot}316 \\ 0{\cdot}316 & 0{\cdot}47 \end{bmatrix} \text{ and that } \left[\frac{\partial D}{\partial b_1}\right]_0 = (0{\cdot}5, 0).$$

Similarly, if $B = \begin{bmatrix} 1 \\ b_2 \end{bmatrix}$ show that $\left[\frac{\partial D}{\partial b_2}\right]_0 = (1{\cdot}96, 0).$

Hence, show that the optimal control vector is

$$u = (-1{\cdot}70 + 0{\cdot}5\delta b_1 + 1{\cdot}96\delta b_2)x_1 - 3{\cdot}16x_2.$$

9–3 Sensitivity of optimal linear control systems to finite variations in parameters

If changes in parameters affecting a mathematical model describing a system are not infinitesimally small then generally nothing can be inferred about the solution to the new problem. However, a fairly fruitful technique is to attempt to find a region within which the parameters may lie so that some property of the solution is invariant. In other words, this property will be completely insensitive to parameter variations provided they lie inside the determined region. This is essentially equivalent to posing an inverse problem and as such is a standard mathematical strategy. One of the few applications to control theory was made by Kalman (1964) who found all quadratic performance indices for which a given linear control law is optimal for a given linear system. There is in addition some Russian work on the 'principle of invariance' and references may be found in a paper by Cruz and Perkins (1966*b*). The insensitivity idea was also used a few years ago in linear programming theory by Gass and Saaty (1955), and Barnett (1962). Indeed, Theorem 8–2–1 is an application of this same principle, where a known Liapunov function for a given system is used to generate an expression for finite changes which can occur without affecting the stability behaviour.

Following Barnett and Storey (1966) let the original system be again described by

$$\dot{x} = A_0 x + B_0 u, \tag{9–3–1}$$

where u is to be chosen so as to minimize the quadratic performance index (9–1–2) with $T = \infty$. The object now is to find what finite changes may occur in A and B (Q and R again remaining fixed) without causing any change in the optimal control matrix

$$D_0 = -R^{-1}B_0'P_0, \tag{9–3–2}$$

where P_0 is the solution of the steady-state Riccati equation (9–1–6). Let the perturbed system be

$$\dot{x} = (A_0 + A_1)x + (B_0 + B_1)u \tag{9–3–3}$$

and let the corresponding solution of (9–1–6) be $P_0 + P_1$. The problem is to find matrices A_1 and B_1 such that the D matrix for (9–3–3) remains at its original value D_0, thus leaving unaltered the optimal feedback control $u = D_0x$. From (9–3–2) this implies

$$R^{-1}(B_0' + B_1')(P_0 + P_1) = R^{-1}B_0'P_0, \tag{9–3–4}$$

or, after simplification,

$$B_1'(P_0 + P_1) + B_0'P_1 = 0. \tag{9–3–5}$$

The Riccati equation for the perturbed system is

$$(P_0 + P_1)(B_0 + B_1)R^{-1}(B_0' + B_1')(P_0 + P_1) - (A_0' + A_1')(P_0 + P_1) - (P_0 + P_1)(A_0 + A_1) - Q = 0,$$

which, using (9–3–4), reduces to

$$P_0B_0R^{-1}B_0'P_0 - (A_0' + A_1')(P_0 + P_1) - (P_0 + P_1)(A_0 + A_1) - Q = 0. \tag{9–3–6}$$

Since P_0 is the solution of the Riccati equation for the original unperturbed system (9–3–6) reduces still further to

$$A_1'(P_0 + P_1) + (P_0 + P_1)A_1 + A_0'P_1 + P_1A_0 = 0. \tag{9–3–7}$$

Thus a solution to the problem is provided by any pair of matrices A_1 and B_1 which satisfy Eqns (9–3–5) and (9–3–7) together with the condition that $P_0 + P_1$ be positive definite. The corresponding minimum value of the performance index is

$$\begin{aligned} J &= [x'(P_0 + P_1)x]_{t=0} \\ &= J_0 + [x'P_1x]_{t=0}, \end{aligned} \tag{9–3–8}$$

where J_0 is the value for the original problem.

It should be noticed that Eqn (9–3–7) is linear in P_1 so that, computationally, the advantage in keeping D fixed is that the new optimization problem thus created depends only upon the solution of a linear matrix equation (in fact a Liapunov matrix equation) for P_1, compared with the quadratic Riccati equation for P_0.

We now derive some expressions for A_1 and B_1 satisfying the required conditions. First write (9–3–7) in the form

$$[P_0A_1 + P_1(A_1 + A_0)] + [P_0A_1 + P_1(A_1 + A_0)]' = 0,$$

whence

$$P_0A_1 + P_1(A_1 + A_0) = S, \tag{9–3–9}$$

where S is a skew-symmetric matrix such that P_1 is symmetric.

Exercise 9–3–1

Show that S in Eqn (9–3–9) satisfies the equation

$$(A_1' + A_0')S + S(A_1 + A_0) = A_0'P_0A_1 - A_1'P_0A_0. \tag{9–3–10}$$

Consider the following special cases:

(i) $A_1 = 0$.

In this case (9–3–7) gives $A_0'P_1 + P_1A_0 = 0$ which, provided A_0 has no characteristic roots with zero real parts (p. 88), has only the trivial solution $P_1 = 0$. Thus if A remains fixed at its original value (9–3–5) cannot be satisfied; that is, any change B_1 in B will cause a change in D.

(ii) $B_1 = 0$.

Eqn (9–3–9) gives [replacing S by $(P_1 + P_0)S_1(P_0 + P_1)$]

$$A_1 = -(P_1 + P_0)^{-1}P_1A_0 + S_1(P_1 + P_0), \qquad (9\text{–}3\text{–}11)$$

where P_1 is any symmetric matrix satisfying (9–3–5):

$$B_0'P_1 = 0 \qquad (9\text{–}3\text{–}12)$$

and the condition that $P_1 + P_0$ be positive definite. Since (9–3–12) represents mn equations for the $\frac{1}{2}n(n+1)$ elements of P_1, a non-trivial solution for P_1 will exist only if $m < \frac{1}{2}(n+1)$. The skew-symmetric matrix in (9–3–11) is entirely arbitrary.

A particular solution of (9–3–12) is, of course, $P_1 = 0$, when

$$A_1 = S_1P_0 \qquad (9\text{–}3\text{–}13)$$

and, from (9–3–8), $J = J_0$. That is, for any A_1 of the form (9–3–13), where S_1 is an arbitrary skew-symmetric matrix, the perturbed system has the same D matrix and same minimum value of the performance index as the original system.

Thus when A is fixed any change in B produces a change in D, but if B is fixed A may vary according to (9–3–13) without altering D. Hence the system can be said to be more sensitive to changes in B than in A.

(iii) $A_1 \neq 0$, $B_1 \neq 0$.

Three cases are considered:

(*a*) If A_1 is given then from (9–3–5) the value of B_1 which leaves D unaltered is

$$B_1 = -(P_1 + P_0)^{-1}P_1B_0, \qquad (9\text{–}3\text{–}14)$$

where $P_1 = (S - P_0A_1)(A_1 + A_0)^{-1}$ and S is the unique solution of (9–3–10). Thus for any given A_1 there is just one corresponding B_1 which leaves D unaltered.

(*b*) If P_1 is chosen to be symmetric and such that $P_0 + P_1$ is positive definite then A_1 is again given by (9–3–11) with S_1 arbitrary, and B_1 is given by (9–3–14) with the known value for P_1. Thus for any particular P_1 there are an infinite number of suitable matrices A_1 but only one matrix B_1.

The advantage of specifying P_1 is that if desired J can be made less than J_0 – for example, by taking P_1 negative definite. That is, systems can be found which are 'better' than the original system (in the sense that J has a smaller value) but still have the same optimal feedback control.

(*c*) An interesting special solution of (9–3–5) and (9–3–7) [obtained by setting $S = 0$ in (9–3–9)] is

$$A_1 = KP_0A_0,\ B_1 = KP_0B_0,\ P_1 = -P_0KP_0(I + KP_0)^{-1}$$

where K is an arbitrary symmetric positive definite matrix. The system (9–3–3) is then

$$\dot{x} = (I + KP_0)(A_0x + B_0u). \tag{9–3–15}$$

It is easily shown that

$$(I + KP_0)'(P_1 + P_0)(I + KP_0) = P_0 + P_0KP_0$$

and

$$(I + KP_0)'P_1(I + KP_0) = -P_0(K + KP_0K)P_0,$$

so that $P_0 + P_1$ is always positive definite and P_1 negative definite. Thus $J \leqslant J_0$ for any choice of K. For example, if K is taken to be diagonal, then as $k_{ii} \to \infty$, $(P_1 + P_0) \to 0$ so that J for the perturbed system can be made as small as desired.

The above cases show that there do exist matrices A_1 and B_1 which leave D unaltered. In this context it is worth noting the following remark by Noton (1965) concerning the design of a lateral autostabilizer for an aircraft: '. . . if the design procedure is carried out for different parts of the flight envelope (when some of the coefficients in (9–3–1) change by a factor of 50) the elements of the D matrix change surprisingly little' [see also Dyer *et al.* (1966)].

A much simpler but related version of the problem of this section is commented on by Rutman and Eppelman (1966). They consider (9–3–1) and discuss what finite changes may occur in A so that a scalar function $y = c'x$ is unaltered. This was extended by Cruz and Perkins (1966*a*) to a vector function $y = Cx$.

Also worth mentioning is a paper by Alexandro (1965) who deals with a situation where the optimal control law for a linear system is non-linear but is kept fixed despite changes in plant parameters. Another approach of some interest has been made by Rissanen (1966), who studies the problem of evaluating the extent to which system parameters may be changed and still guarantee that the value of the performance index will remain within a specified limit.

If A is kept fixed and B allowed to vary then it is easy to show from (9–3–6) that Q must be changed if D is not to alter, and the change in Q uniquely determines the change necessary in D. However, this means that the optimization problem is essentially a different one, since the performance criterion has been altered. The problem of investigating the effect of changes in the performance index is an important one, as yet unsolved, although Tyler and Tuteur (1966) have made an interesting contribution.

Exercise 9–3–2 [Barnett & Storey (1966)]
Using the matrices of Exercise 9–2–2 show that

$$A = \begin{bmatrix} -1 + 0{\cdot}316\alpha & 0{\cdot}85\alpha \\ 1 - 0{\cdot}17\alpha & -0{\cdot}316\alpha \end{bmatrix}, \quad B = B_0$$

(α arbitrary) leaves the optimal control law unaltered. Similarly, use (9–3–15) to obtain

$$A = \begin{bmatrix} -1 & 0{\cdot}15k_1 \\ 1 & 0{\cdot}53k_2 \end{bmatrix}, \quad B = \begin{bmatrix} 1 + 0{\cdot}17k_1 \\ 0{\cdot}316k_2 \end{bmatrix}, \quad (k_1, k_2 \geqslant 0).$$

9–4 Insensitivity of general optimal control systems

The idea of the previous section may be generalized for a class of optimal control problems [Barnett (1968)]. Let the original system be described by

$$\dot{x} = f(x, u) \qquad [x(0) \text{ specified}], \tag{9–4–1}$$

the control vector $u = (u_1, u_2 \ldots, u_m)'$ being chosen so as to minimize a performance index

$$F = \int_0^T L(x, u)\, dt, \tag{9–4–2}$$

where the final time T and $x(T)$ are fixed. As usual the hamiltonian is defined by

$$H = \sum_{i=1}^{n} \psi_i f_i - L \tag{9–4–3}$$

where the functions $\psi_i(x, t)$ satisfy the adjoint equations $\dot{\psi}_i = -\partial H/\partial x_i$. Define also

$$V(x, t) = \min_u \int_t^T L(x, u)\, dt \tag{9–4–4}$$

and it then follows that $\psi_i = -\partial V/\partial x_i$. The Pontryagin maximum principle requires that H as defined by (9–4–3) be a minimum with respect to u, leading to the optimal control $u^0(x, \psi)$ as a function of state and adjoint variables. For further details, including restrictions on f and L, etc., a standard reference such as Rozonoer (1959) should be consulted.

The perturbed system is taken to be of the form

$$\dot{x} = f(x, u) + k(x, t)$$

with the same initial conditions and performance index as for (9–4–1). The problem is to determine the vector function k so that (*a*) the optimal control vector, as a function of state and adjoint vectors, and (*b*) $V(x, t)$ as defined by (9–4–4) both remain unchanged. Thus the problem is a restricted version of the linear system case, since in particular (*b*) implies that the minimum value of the performance index is also unaltered.

Further analytical details of the problem and its solution are outside the scope of this text; see instead Barnett (1968). The solution turns out to be determined by the equation

$$K\psi = -\Psi k \tag{9–4–5}$$

where the $n \times n$ matrices K, Ψ are given by $K = (\partial k_j/\partial x_i)$ and $\Psi = (\partial \psi_i/\partial x_j)$. Eqn (9–4–5) reduces to

$$\nabla(k'\psi) = 0 \tag{9–4–6}$$

which implies that $k'\psi$ is independent of x.

For example, in the constant linear case of the previous section, $\psi = -2Px$ so that $\Psi = -2P$, and (9–4–5) gives $k = SPx$ where S is an arbitrary skew-symmetric matrix. This agrees with (9–3–13) but it is easily verified that the elements of S may now be functions of x and t. That is, for any function $k(x, t) = S(x, t)Px$, $k(0, t) = 0$, the system $\dot{x} = Ax + Bu + k$ has the same optimal control as the original system (9–3–1).

The foregoing example suggests that a solution of (9–4–5) may be

$$k(x, t) = S(x, t)\psi(x, t), \tag{9–4–7}$$

and it is easy to see from (9–4–6) that this is indeed the case. The original system is completely insensitive, according to the stated requirements, to any perturbing function of the form (9–4–7).

Further examples are discussed in Barnett (1968). In one of these it turns out that if the performance index for a constant linear system contains terms in the state variables of degree greater than two [Bass and Webber (1966)] then any change in A alone causes a change in the optimal feedback control, showing that the choice of performance criterion affects the sensitivity of the optimal control.

For a similar, although independent, approach to the same problem see McClamroch and Aggarwal (1967).

9–5 Linear system functionals

Consider once again an asymptotically stable linear system described by

$$\dot{x} = Ax, \tag{9–5–1}$$

where A is an $n \times n$ real matrix. It is of considerable importance to have some quantitative measure of the system transient response without explicit determination of the solution $x(t)$ of (9–5–1). One of the simplest such measures is the functional

$$\int_0^\infty x' Kx dt, \tag{9–5–2}$$

where K is a constant symmetric positive definite matrix. It is easy to show [Bellman (1960)] that the value of (9–5–2) is just $-x'(0)\,Lx(0)$, where L is the solution of the Liapunov matrix equation $A'L + LA = K$. An easy extension was made to a wider class of functionals by Macfarlane (1963).

Exercise 9–5–1

Show that $\frac{d}{dt}(x'Lx) = x'Kx$, where $x(t)$ is the solution of (9–5–1). Hence show that $\int_0^t x'Kx\ dt = x'(t)Lx(t) - x'(0)Lx(0)$ and thus verify the above result.

Macfarlane (1965) has developed a method whereby a very large class of functionals of system response may be evaluated without evaluation of $x(t)$. He introduces a generalized state vector $x_{(m)}$ with components of the form

$$\frac{m!}{a!b!\cdots h!} x_1^a x_2^b \cdots x_n^h$$

where $a + b + \cdots + h = m$, all possible combinations occurring in an order to be defined below, so that $x_{(m)}$ has

$$\binom{n+m-1}{m} = \frac{n(n+1)\cdots(n+m-1)}{m!} \equiv N(m, n)$$

components. Associated with $x_{(m)}$ is the so-called *general functional matrix* $M(m, n)$ which satisfies

$$\frac{d}{dt} x_{(m)} = Mx_{(m)}. \tag{9–5–3}$$

This $N(m, n) \times N(m, n)$ matrix M is uniquely defined by the relationships

$$q = M'p \tag{9–5–4}$$

and

$$\frac{dP(x)}{dt} = Q(x), \tag{9–5–5}$$

where $Q(x)$ and $P(x)$ are time independent homogeneous algebraic forms of degree m in x and

$$P(x) = p'x_{(m)}, \qquad Q(x) = q'x_{(m)}, \tag{9–5–6}$$

p and q being vectors obtained by arranging the coefficients of $P(x)$ and $Q(x)$ in a defined order. In particular if $p_{\alpha\beta\gamma\ldots\mu}$ is a general element of p then each sequence of suffixes containing k digits $\alpha\beta\gamma\cdots\mu$ is formed from a sequence containing $k - 1$ digits by choosing α so that $\alpha \leqslant \beta \leqslant \gamma \cdots \leqslant \mu$. For example, when $n = 3$, $m = 2$ the sequence of suffixes ij for the elements p_{ij} and q_{ij} is (11, 12, 22, 13, 23, 33) and $x'_{(2)} = (x_1^2, 2x_1x_2, x_2^2, 2x_1x_3, 2x_2x_3, x_3^2)$; when $n = 3$, $m = 3$ the sequence of suffixes ijk for the elements p_{ijk} and q_{ijk} is (111, 112, 122, 222, 113, 123, 223, 133, 233, 333) and

$$x'_{(3)} = (x_1^3, 3x_1^2x_2, 3x_1x_2^2, x_2^3, 3x_1^2x_3, 6x_1x_2x_3, 3x_2^2x_3, 3x_1x_3^2, 3x_2x_3^2, x_3^3).$$

It follows at once from Eqn (9–5–5) that

$$\begin{aligned} \int_0^\infty Q(x)\,dt &= -P'[x(0)] \\ &= -p'x_{(m)}(0) \qquad \text{using (9–5–6)} \\ &= -q'M^{-1}x_{(m)}(0) \quad \text{using (9–5–4),} \end{aligned}$$

so that the inverse of the general functional matrix needs to be calculated. This is also necessary for more general functionals of the form

$$\int_0^\infty f(t)\,Rx_{(m)}\,dt$$

(where R is a matrix of arbitrary constants and $f(t)$ a time-weighting function) for which Macfarlane gives an expression in terms of M and $x_m(0)$. The direct method of obtaining M is simply to write, using (9–5–1) and (9–5–5),

$$x'A'\,\nabla\,P(x) = Q(x) \tag{9–5–7}$$

and then solve for the elements of M between Eqns (9–5–4) and (9–5–7). Macfarlane gives some examples for cases up to $n = 5$, but the work involved is tedious and soon becomes very lengthy [for example, $N(3, 4) = 20$, $N(3, 5) = 56$]. However, a computer program for obtaining M has been developed by Undrill (1967) and this has been found to work well in practice. More recently Macfarlane (1969) has shown how Kronecker products and sums can be used to simplify his method.

A different, algebraic approach was developed by Barnett and Storey (1967). There we showed how $M(m, n)$ can be determined iteratively from $M(2, n)$ using

$$A \otimes I_{N(m-1,n)} + I_n \otimes M(m-1, n).$$

The matrix $M(2, n)$ is just the matrix B of (6–1–7) and may be obtained as mentioned in Chapter 6.

We shall go into no further details here but simply refer interested readers to our original paper; it should be pointed out, however, that a numerical comparison of the two methods for computing $M(m, n)$ [Barnett, Ford, and Storey (1968)] established that Undrill's approach is in fact the more efficient. In this last-mentioned paper application was also made to linear control systems with higher than second-degree terms in the performance index [Bass and Webber (1966)].

Exercise 9–5–2 [Macfarlane (1963)]
Let L_1 be the solution of $A'L_1 + L_1A = K$. Show that

$$\frac{d}{dt}(t\,x'\,L_1\,x) = x'\,L_1x + t\,x'\,Kx.$$

Hence by integrating both sides and using the result of Exercise 9–5–1, show that

$$\int_0^\infty t\,x'\,K\,x\,dt = (-1)^2\,x'(0)\,L_2x(0)$$

where $A'\,L_2 + L_2\,A = L_1$.

A continuation of this argument gives

$$\int_0^\infty t^r\,x'\,Kx\,dt = (-1)^{r+1}\,r!\,x'(0)\,L_{r+1}\,x(0) \quad \text{where}$$

$A'\,L_{s+1} + L_{s+1}\,A = L_s, \quad s = 1, 2, \ldots, r.$

Discussion and further reading

The object of this final chapter has been to sketch some applications of the previous results and methods to topics in modern control theory. Details have been deliberately kept as brief as possible, the object being mainly to whet the reader's appetite so that he may be encouraged to explore further into the wide range of problems which exist in the field. To help achieve this aim a considerable number of references is given; others, together with suggestions for future research, are given by Barnett (1967).

References

ALEXANDRO, F. J. (1965), Compensation of plant variations in optimal control systems, *I.E.E.E. Trans. Aut. Control*, **AC–10,** 275–280.

ATHANS, M. (1966), The status of optimal control theory and applications for deterministic systems, *I.E.E.E. Trans. Aut. Control*, **AC–11,** 580–596.

BARNETT, S. (1962), Stability of the solution to a linear programming problem, *Operat. Res. Quart*., **13,** 219–228.

BARNETT, S. (1966), Sensitivity of optimal linear systems to small variations in parameters, *Int. J. Control*, **4,** 41–48.

BARNETT, S. (1967), Solution of some algebraic problems arising in the theory of stability and sensitivity of systems, with particular reference to the Liapunov matrix equation, Ph.D. Thesis, Loughborough University of Technology.

BARNETT, S. (1968), *Insensitivity of control systems*, presented at the 3rd United Kingdom Automation Council Convention, University of Leicester, Reprinted in *Int. J. Control*, **10,** 665–675 (1969).

BARNETT, S. and C. STOREY (1966) Insensitivity of optimal linear control systems to persistent changes in parameters, *Int. J. Control*, **4,** 179–184.

BARNETT, S. and C. STOREY (1967), On the general-functional matrix for a linear system, *I.E.E.E. Trans. Aut. Control*, **AC–12,** 436–438.

BARNETT, S., B. FORD and C. STOREY (1968), *Comparison of several methods for calculating the general functional matrix for linear systems*, presented at the 3rd United Kingdom Automation Council Convention, University of Leicester. Reprinted in *Int. J. Control*, **10,** 677–686 (1969).

BASS, R. W. and R. F. WEBBER (1966), Optimal non-linear feedback control derived from quartic and higher-order performance criteria, *I.E.E.E. Trans. Aut. Control*, **AC–11,** 448–454.

BELLMAN, R. (1960), *Introduction to matrix analysis*, McGraw-Hill, New York.

BLACKBURN, T. R. and J. C. BIDWELL (1968), Some numerical aspects of control engineering computations, *Preprints Joint Automatic Control Conf.*

CRUZ, J. B. and W. R. PERKINS (1966*a*), Conditions for signal and parameter invariance in dynamical systems, *I.E.E.E. Trans. Aut. Control*, **AC–11,** 614–615.

CRUZ, J. B. and W. R. PERKINS (1966*b*), Criteria for system sensitivity to parameter variations, *Proc. I.F.A.C. Conference*, London.

DYER, P., A. R. M. NOTON and D. RUTHERFORD (1966), The application of dynamic programming to the design of invariant auto-stabilisers, *J. Roy. Aero. Soc.*, **70,** 469–476.

FRIEDLAND, B. (1967), On solutions of the Riccati equation in optimization problems, *I.E.E.E. Trans. Aut. Control*, **AC–12,** 303–304.

GASS, S. I. and T. L. SAATY (1955), The computational algorithm for the parametric objective function, *Naval Res. Logistics Quart.*, **2,** 39–46.

KALMAN, R. E. (1964), When is a linear control system optimal?, *Trans. A.S.M.E.*, Series D, **86,** 51–60.

KALMAN, R. E. (1960), Contributions to the theory of optimal control, *Bol. Soc. Mat. Mexicana*, **5,** 102–119.

KOKOTOVIC, P. V. and R. S. RUTMAN (1965), Sensitivity of automatic control systems, *Automat. Rem. Control*, **26,** 727–749.

MACFARLANE, A. G. J. (1969), Functional matrix theory for the general linear electrical network – Part 5, *Proc. I.E.E.*, **116,** 1745–1747.

MACFARLANE, A. G. J. (1963), The calculation of functionals of the time and frequency response of a linear constant coefficient dynamical system, *Q. J. Mech. Appl. Math.*, **16,** 259–271.

MACFARLANE, A. G. J. (1965), Functional-matrix theory for the general linear electrical network – Part 2, *Proc. I.E.E.*, **112,** 763–770.

MCCLAMROCH, N. N. and J. K. AGGARWAL (1967), On equivalent systems in optimal control and stability theory, *I.E.E.E. Trans. Aut. Control*, **AC–12,** 333.

MERRIAM, C. W. III (1964), *Optimisation theory and the design of feedback control systems*, McGraw-Hill, New York.

NOTON, A. R. M. (1965), *Introduction to variational methods in control engineering*, Pergamon Press, Oxford.

PAGUREK, B. (1965), Sensitivity of the performance of optimal control systems to plant parameter variations, *Int. J. Control*, **1,** 33–45.

PORTER, W. A. (1967), On the matrix Riccati equation, *I.E.E.E. Trans. Aut. Control*, **AC–12,** 746–749.

POTTER, J. E. (1966), Matrix quadratic solutions, *SIAM J. Appl. Math.*, 496–501.

REID, W. T. (1959), Solutions of a Riccati matrix differential equation as functions of initial values, *J. Math. and Mech.*, **8,** 221–230.

RISSANEN, J. J. (1966), Performance deterioration of optimum systems, *I.E.E.E. Trans. Aut. Control*, **AC–11,** 530–532.

ROZONOER, L. I. (1959), The maximum principle of L. S. Pontryagin in optimal system theory, *Automat. Rem. Control*, **20,** 1288–1302; 1405–1421; 1517–1532.

RUTMAN, R. and M. EPPELMAN (1966), On necessary and sufficient conditions of parametric invariance, *Sensitivity methods in control theory*, ed. L. Radanovic, Pergamon Press, Oxford, 290–293.

SOBRAL, M. (1968), Sensitivity in optimal control systems, *Proc. I.E.E.E.*, **56,** 1644–1652.

TOMOVIC, R. (1963), *Sensitivity analysis of dynamic systems*, McGraw-Hill, New York.

TYLER, J. S. and F. B. TUTEUR (1966), The use of a quadratic performance index to design multivariable control systems, *I.E.E.E. Trans. Aut. Control*, **AC–11,** 84–92.

UNDRILL, J. M. (1967), On the construction of Macfarlane's generalised functional matrix, *Proc. I.E.E.E.*, **114,** 166–167.

Author Index

Subject Index